Osamu Sudoh (Ed.)

Digital Economy and Social Design

Osamu Sudoh (Ed.)

# Digital Economy and Social Design

Springer

Osamu Sudoh, Dr.
Professor
Graduate School of Interdisciplinary Informatics
The University of Tokyo
7-3-1 Hongo, Bunkyo-ku
Tokyo 113-0033, Japan

Library of Congress Control Number: 2005922604

ISBN 4-431-25467-6 Springer-Verlag Tokyo Berlin Heidelberg New York



Springer is a part of Springer Science+Business Media
springeronline.com

Printed in Japan

Typesetting: Camera-ready by the editor.
Printing and binding: Nikkei Printing, Japan

Printed on acid-free paper

# Preface

The advent of the digital economy has the potential to dramatically change the conventional interrelationships among individuals, enterprises, and society, and to make a considerable qualitative difference to major socioeconomic systems. The expanded applications of information technology have actually lowered the significance of boundaries between organizations and between countries, and have intensified competition among businesses and institutions. In addition, a paradigm shift of unprecedented dynamism is emerging in a very broad range of areas, from the world economic order to personal lifestyles. In the business world, for instance, more cross-sector collaborative relationships are being established and different business models are being integrated. In government, citizens' participation is growing and administrative functions are changing. Communications between different cultures via the Internet are expanding. In industrial, economic, institutional, educational, cultural, and many other spheres, interactions among different communities or countries are gathering pace and structural changes are accelerating at national and transnational levels.

These structural changes have just begun recently. Many are not fully in place. But there can be little doubt that to achieve vigorous socioeconomic development in the twenty-first century, people will have to aggressively use information technology to boost innovation and to organically link the results of that innovation to solutions to global environmental issues and social challenges such as the opportunity divide.

The digital economy has in fact been posting steady growth, even since the bursting of the new economy bubble. We now need dispassionate analyses of the impact of information and telecommunications technologies on socioeconomic systems and the development of a practical institutional design. In this sense, the compilation of such analyses into a single book is timely.

This book is based on an international conference on the establishment of the digital economy and the evolution of socioeconomic systems. The conference was held on February 27, 2003, at the University of Tokyo. It was organized by Research Project A06, entitled "Researching the Emerging Global Information Economy and Designing a New Social System," which I had the honor of leading. This project has been designated as one of the "Informatics Studies for Building the Foundation of Deepening Information Technology," and as such is eligible for subsidies from the Ministry of Education, Culture, Sports, Science, and Technology to cover scientific research costs. The conference was cohosted by the Institute of Socio-information and Communication Studies at the University of Tokyo. Benefiting from the participation of an audience with a broad interest in the economy, business administration, and many other issues, the conference enjoyed a lively discussion. This book is a substantially rewritten version of the conference proceedings, adopted after strict peer review by anonymous reviewers and reflecting both the discussions that took place at the conference and subsequent trends in economic society.

The book consists of three parts, each featuring papers mainly based on theoretical research and papers centered on empirical research. Each part examines one topic, as follows:

Part I. The Macro and Micro Economic Aspect
Chapters 1 to 3 examine the impact of informatization on the overall socioeconomic system, what will be necessary to make effective use of information and knowledge to sustain the development of society and the economy, whether or not new technological innovation will aggravate inequality, and what skills, techniques, and institutional design will be required to ensure that different individuals and groups can broadly reap the benefits of the digital economy.

Part II. The Nature of Competition
Chapters 4 to 6 focus on the functions performed in the network society with its constantly changing organizational boundaries by the mechanism of credit rating and selection in the market, how businesses compete with one another in the digital economic environment, and how this competition differs from the existing competition.

Part III. Structural Changes
Chapters 7 to 9 assess how the digital economic environment changes the structure and management of enterprises and the government, how businesses, the government, and citizens will change after the information society emerges, and what forms of governance should be adopted.

The digital economy is by no means an inexorable and deterministic phenomenon. Rather, it is an ever-changing social framework. It is of significance to researchers. The object of their study is so complex that it often involves complex relationships. We are responsible for taking advantage of the opportunities opened up by the digital economy and for turning those opportunities into things that reflect our values and goals. This publication will have achieved its goal if the exploration and questions it raises serve as a signpost for other research projects and policy making.

I would like to conclude this preface by expressing my sincere gratitude to all those who offered support and cooperation in the publication of this book. In particular, I would like to give special thanks to Ms. Reiko Gotoh for her outstanding contribution to the planning and compilation of this publication, serving as a research partner in the project "Researching the Emerging Global Information Economy and Designing a New Social System" and also as secretary of the international symposium "The Establishment of the Digital Economy and the Evolving Socioeconomic Systems." I am also deeply indebted to Dr. Yuichiro Anzai, a professor of Keio University and field representative for "informatics" and others who willingly agreed to proofread for us, to the editorial staff at Springer-Verlag, Tokyo, for their great help with the publication of this book, and to the Ministry of Education, Culture, Sports, Science, and Technology for financially supporting our research project.

Osamu Sudoh

February 2005

# Contents

# Part I: The Macro and Micro Economic Aspect

# 1 The Knowledge Network in the Digital Economy and Sustainable Development

Osamu Sudoh

Graduate School of Interdisciplinary Informatics, The University of Tokyo, 7-3-1, Hongo, Bunkyo-ku, Tokyo 113-0033, Japan, sudoh@iii.u-tokyo.ac.jp

**Summary**

Today, the global environment represents the most crucial issues we will have to face. Based on the principles of sustainable development, local residents, the government, local authorities and companies must cooperate under the banner of participatory local democracy, with nonprofit organizations (NPOs) playing a key role as junction points, to work toward the environmentally-friendly development of local communities. These local communities then need to be linked on a global scale to create a worldwide network for the development of a society that is geared towards preserving its environment. On this chapter, we consider on the interactive relationship between the market economy and technological development, and the evolution of information technology (IT) and the digital economy. Next, based on these trends, we will examine the relationship between the formation of a networked society and regional development. We will then go on to consider how IT and information networks can contribute to a new form of social development that is environmentally friendly.

**Key Words** Digital economy, Web-based community, Sustainable development, e-Government, Knowledge network

## 1.1 Introduction

Looking into the twenty-first century, the global environment represents the most crucial issues we will have to face. Yet, it is no easy task to come up with a basic design for the sustainable development of mankind, or to implement this basic design on more concrete levels. In reality, there is a complex web of factors, with each country facing its own specific problems that reflect its unique history, creating numerous conflicts of interest.

In the long term, as the threat of global environmental destruction grows ever greater, global networks that link regions to the world may prove a powerful force in the formation of new societal standards that will triumph over these nationalistic interests. Based on the principles of sustainable development, local residents, the government, local authorities and companies must cooperate under the banner of participatory local democracy, with nonprofit organizations (NPOs) playing a key role as junction points, to work toward the environmentally-friendly development of local communities. These local communities then need to be linked on a global scale to create a worldwide network for the development of a society that is geared towards preserving its environment.

This chapter begins with a look at the interactive relationship between the market economy and technological development, and the evolution of information technology (IT) and the digital economy. Next, based on these trends, we will examine the relationship between the formation of a networked society and regional development. We will then go on to consider how IT and information networks can contribute to a new form of social development that is environmentally friendly. More specifically, this chapter takes the case of the Zero-Emissions Research Initiative, and examines how the expansion of the digital revolution should be used to promote sustainable development.

## 1.2 The Market Economy and Industrial Networks

### 1.2.1 The Interactive Relationship Between the Market Economy and Technology

After World War II, a model for economic development that centered on mass production and mass consumption took hold in developing countries, and the world witnessed widespread environmental destruction. The causes of this massive degredation of the environment can be broadly divided into two. The first is the large-scale destruction of the environment that was brought on by the materialistic, wealth-oriented economic systems built by advanced nations, or namely systems of mass production and mass consumption, and the mass disposal that accompanied them. The second is the deterioration of natural resources, such as forests, water and soil, caused by poverty in developing countries, foreign exchange shortages and specializations in commercial crop production, the amassing of land by certain privileged classes, and rapid population increases. These two forms of environmental destruction are connected in complex and interactive ways.

Throughout the twentieth century, the words "progress" and "development" in most cases were used to mean "economic progress" and "economic development," and it is no exaggeration to say that the natural environment was seen as a given, and in this short-sighted view, the expansion of monetary wealth and material benefits took priority. Moreover, with the exception of localized water and air pollution, little consideration was given to the destructive effects of economic activities on nature until global environmental problems became much more serious.

Yet, economic activities dramatically boosted productivity (and our power over nature), and as a result, economic activities and technology came to have a massive impact on the natural environment. The assumption that the natural environment was a given for economic activity had by this stage become a thing of the past. Let's begin by looking at the relationship between the market economy and technological development.

Here we shall look at the interactive relationship between technology and the market, based on the observations of Michael J. Piore and Charles F. Sable (Piore and Sabel 1984:38–48).

Piore and Sable have made some very interesting observations about the relationship between technological progress in America, England and France in the early nineteenth century and domestic markets. According to their findings, early nineteenth century America suffered from a shortage of highly-skilled workers, and unlike Europe, did not have guilds that restricted the organization of production. There were also numerous affluent independent farmers who had a certain degree of purchasing power. Given these market conditions, entrepreneurs needed to organize production processes to reduce their dependence on labor, and this in turn required the introduction of labor-saving machinery. What's more, the independent farmers, who comprised the majority of the consumer base, willingly bought up the standardized products manufactured using this machinery. By then, the foundations had already been laid for the development and introduction of mass-production technology in America.

Meanwhile in France, while French farmers owned the rights to control their own land, as did farmers in America, unlike their American counterparts, the majority of French farmers owned small plots of land and their livelihoods were not integrated with the market economy. Demand for manufactured goods thus came only from the aristocracy, bourgeois merchants, bureaucrats and the few affluent farmers.

Yet regional markets were very diverse in nature, causing segmentation in the domestic market. Guilds existed in France and although they did not have very strong regulatory powers, the highly-skilled workforce and sophisticated techniques cultivated by the guilds were used to develop new manufactured goods and manufacturing processes. These market conditions encouraged the French economy to specialize in producing a wide variety of high-quality goods. As such, there was very little incentive in France to develop and implement mass-production technology at this time.

In early nineteenth century England fell somewhere between the two extremes of America and France. Land enclosure moves in England created a sharp division amongst independent farmers, with the majority of farmers losing control of their land and migrating to urban centers as workers. This generated a huge market for low-cost consumer goods in urban areas, while there was also a market for low-cost consumer goods in the colonies. On the supply side, merchandise production was beginning to be organized by those who had evaded the guild or state regulations. Meanwhile industrialization flowed over into rural areas, increasing the population density in rural areas, and as a result consumer preferences in regional areas became more stable. As England had an abundance of both skilled and unskilled workers, entrepreneurs did not need to come up with labor-saving initiatives. Consequently, while England developed mass-production systems in the cotton textiles, food processing and brewing industries, market conditions did not facilitate the aggressive development and introduction of non-labor intensive standardized mass production technology that was seen in the United States.

In this way, the "market" had a very different social function in each of these countries because of their differing economic climates. For this reason, the path of technological progress in each of these countries differs greatly, a reflection of their individual market peculiarities. Therefore, when examining the relationship between environmental problems and economic growth, we cannot consider the two in binary opposition, but must rather fully understand the interactive relationships between the underlying market characteristics, social systems, technology, governments, and social customs.

### 1.2.2 The Development of Industrial Networks

In nineteenth century Europe and America, many industrial areas achieved economic growth after organizing local industry networks. Some classic examples of this include the silk textiles manufacturing industry in Lyon, the cutlery and specialty steel manufacturing industries in Solingen, calico manufacturing in Alsace, cotton textiles manufacturing in Philadelphia, and specialty steel manufacturing in St. Etienne. Piore and Sabel believe that there were specific mechanisms that facilitated economic growth in these areas (Piore and Sabel 1984:19-48).

Below are some important commonalities between these advanced industrial areas at the time, as identified by Piore and Sabel.

First, these industrial areas had found ways to readily adjust the combination of technology and capital equipment in the industrial system to adapt flexibly to fluctuations in demand. In Lyon and Solingen for example, small specialty producers joined together to form local industry networks. But there were no fixed relationships between prime contractors and sub-contractors, and the relationships between companies, or more accurately, the relationships between work places, were flexible and dynamic, with companies sometimes being the prime contractor and then at other times the sub-contractor. Therefore, the relationships between work places were organized and re-organized as needed to respond to fluctuations in demand, and local industry networks were formed that were both diverse and flexible.

Second, the work places and factories in these industrial areas placed great importance on innovation and worked to improve productivity, while they also had the capacity to flexibly cater to demand fluctuations and the ability to create new markets. In addition, producers made concerted efforts to adapt and apply technology developed for diversified applications. This meant that the development and use of advanced technology within these areas was fiercely competitive.

In Lyon for example, the jacquard loom was developed in the first half of the 19th century. With the introduction of the jacquard loom, punch cards were used to control textile patterns, increasing precision, and enabling the textile patterns to be changed simply by replacing the punch card. Using this jacquard loom, the Lyon textiles industry could successfully adapt to fluctuating demand. Meanwhile, research into steam-powered engines and gasoline engines was underway in Solingen.

Third, to facilitate stable development of local industry, the local industry networks established systems that were very similar in nature to local co-operatives. For example, in Solingen and St. Etienne, vocational training schools were set up for the general benefit of the local community, and training and retraining in skills and techniques was systematically conducted. This provided a foundation for local communities, and enabled producers to cope with changes in technology and fluctuating demand appropriately. It also actively encouraged workers to develop expertise, scientific knowledge and creativity, ensuring continued innovation. The systems devised ways of acquiring not just techniques and skills, but broad-based information and knowledge, they maintained cooperative relationships between workers and management, and endeavored to restrict low-price competition that generates wage cuts and declining quality. In other words, they maintained economic order in the local community, and applied harsh entry restrictions to prevent excessive competition, ensuring that the welfare of members of the community was kept above a certain level.

In short, the many companies, or rather work places, in each industrial area, (1) became highly specialized, and ensured that production could be adapted to demand, and therefore (2) there was an invisible collective impetus throughout the community which encouraged the rapid adoption of technological developments and the latest technology in industry, thereby promoting competition. Meanwhile (3) price and wage competition, which would impede the adoption of technological innovation or the latest technologies in industry, was restricted through labor-management agreements, and market entry was restricted to ensure that the local community's high-quality human resources were regenerated. In this way, they responded to market fluctuations whilst maintaining a balance between cooperation and competition, and organized the market place.

So local industry networks were formed in nineteenth century Europe, and there were many successful examples of economic development in which competitive and regulatory tools were used with harmony between market-oriented and non-market-oriented systems.

Subsequently, the 1920s saw the emergence of mass production systems used by the corporate giants. The plants constructed by American automaker Ford Motor Company in the period between the two great wars represented the new paradigm. The mass production system developed by Ford, otherwise known as the Ford System, was subsequently adopted around the world. After World War II, improvements to the social welfare system for unemployment benefits, pensions and health insurance relieved anxieties over unemployment, retirement and illness, while the introduction of installment sales generated a mass consumption base that would absorb this mass production. A development model based on mass production and mass consumption and known as Fordism, was established, and from the late nineteenth century through to the mid twentieth century, we saw the gradual disappearance of the development model that had centered on globally dispersed local industry networks.

Meanwhile, advanced counties like the United States, Western European countries and Japan invested large volumes of energy and resources to achieve remarkable economic growth. Under Fordism, economic growth is the major prerequisite for all manner of social functions, and innovation and mass production-mass consumption are essential elements in achieving economic growth. Even with an inequitable distribution of wealth, economic growth and job security mean that corporations can reap greater profits and workers can earn higher wages. Additionally, the consumption levels of the majority of the population is dependent on their income from wages, so the government now had to give priority to expansionary economic policies to sustain and generate employment, otherwise they would lose the support of the populace. Not adopting policies for economic expansion would result in less tax revenue, which would harm fiscal resilience.

Any environmental considerations at this point would have been little more than stopgap measures. Resources and energy were consumed in vast quantities, and corporate profit-motivated innovations and initiatives meant that products that could still be used were simply discarded. Awareness of resource utilization and the environment started to grow in advanced countries from the late 1960s and 1970s, sparking a re-examination of the mass production-mass consumption concept. Yet the economic growth that had been sustained for thirty years since the end of the war was beginning to decline. Starting the 1960s, the growth rate of labor productivity went into a gradual decline, damaging profitability, and the two oil shocks of the 1970s brought the mass production-mass consumption economy to its knees. As the economic crisis deepened, social awareness was drawn away from resources and the environment. Interest shifted away from the importance of environmental issues to focus on economic problems, more specifically, unemployment and the search for new economic growth models.

Given these worldwide conditions, many firms, especially Japanese firms, channeled their technological development into process innovation, cut back on energy and materials per unit output and endeavored to increase labor productivity by reducing the volume of employment per unit output. Economic growth, albeit sluggish, did result in an expansion in output and sales volumes, but ultimately overall energy and natural resource consumption also increased.

Then the 1980s gave rise to the neo-liberalism movement in the United States and Britain, where the focus turned toward market competition, with the adoption of supply-side strategies being favored over demand-side strategies. Yet in that time the development of micro electronics had prompted the rapid adoption of information and communications technology in industry to cut costs and generate new markets. This was accompanied by increased networking between companies and markets. Piore and Sabel noted in the mid-1980s signs of a resurgence in network-based industry organizations, with coordinating roles as summarized in the three points cited earlier (Piore and Sabel 1984:265-272). So what happened to inter-company and market networks from the 1980s onwards?

Today there are countless semi-horizontal and semi-vertical inter-company networks that are based on information and communications technology, and companies are actively expanding their involvement in technology transfers, joint research, joint ventures and a variety of strategic alliances. These inter-company networks have been developing gradually since the 1970s, but an examination of growth patterns in the 1980s has identified two specific patterns.

The first pattern was the case in which inter-company networks spanned a number of countries. This pattern closely resembled that of semi-vertical networks. Strategically critical sectors like planning, research and development and finance were concentrated in major cities throughout North America, Asia and Western Europe, and manufacturing and sales sectors operated in various forms, either as subsidiaries or joint ventures with majority ownership by the parent company, or through partnerships with other companies, and were scattered around the globe. Looking at companies that fit this pattern, we see that they have implemented comprehensive labor cost-cutting measures (with insecure employment and the transfer of production bases to developing countries), and have further separated their planning and operations divisions. In many cases, manufacturing processes with the greatest potential for causing environmental pollution have been transferred to developing nations with lax environmental regulations.

In the second pattern, inter-company networks were formed within a confined area, national economic zone or regional economic zone. Within specific economic zones, a major corporation would head up a semi-vertical network or a semi-horizontal network consisting of a number of companies. There were many examples of this network pattern in Germany, Sweden, Northern Italy and Japan. Under this pattern we additionally saw the formation of multi-faceted networks that also counted universities and local governments as participants. In Germany, Sweden, Denmark and other northern European nations, and especially in Japan, labor and management cooperated to improve quality and productivity, and cooperative relationships between companies, universities and local governments were formed and strengthened. Generally speaking, there was a very strong focus on both environmental protection and economic growth throughout Germany and Northern Europe, and this was probably not unrelated to the cooperative relationships that existed in these local industry networks.

These two inter-company networking patterns emerged with the increasing use of advanced technology and information networking in the 1980s. If we look at the increased inter-company networking since the mid 1990s when the Internet entered the commercial arena, there seems to have been greater interaction and cross penetration between the two patterns. Networking today is characterized by the formation of diverse and complex global inter-company relationships through joint ventures, technical cooperation, original equipment manufacturing (OEM: products made under a partner's brand), and joint R&D. Companies consolidate their mutual advantages to promote innovation and to diversify their costs and risks, so as to be able to flexibly cater to a number of markets each with their own regional peculiarities. This kind of networking differs from the two networking patterns prevalent in the 1980s, but does incorporate elements of both.

## 1.3 The Emerging Digital Economy and the Paradigm Shift of the Market Structure

### 1.3.1 The Macroeconomic Significance of the Digital Economy

Since the mid-1990s there has been a push to create a massive global market built around the Internet. This is, in other words, the digital economy. America, Europe and Japan have been aggressively promoting the expansion of the digital economy, and as a result increasingly complex and diverse relationships are being formed on a global scale.

Looking back, up until around 1998 there was obvious skepticism about the effects of IT and the Internet on the economy. Christopher Freeman's comments (Freeman, C. 1992) are noteworthy in this regard. He pointed out that in most major OECD countries, the growth rate of labor productivity was exceptionally high in electronics industries such as computers and electronics components, and there was a significant boost in capital productivity as well. Yet, in most countries, the energy- and resource-intensive mass production paradigm still held the greatest weight across industry as a whole, which was dominated by corporate organizations, industrial regulations, social systems and infrastructure investment designed for the mass-production paradigm. The new IT-based paradigm was not consistently present throughout the economy, and any spill-over effects were highly irregular. Even if individual companies were to actively introduce IT, corporate structures, social systems, industrial regulations and infrastructure were still geared to the old mass-production paradigm, which meant that the revolutionary capabilities of IT could not fully permeate the economy, and any impact on society was limited to specific areas.

However, information network upgrades, corporate organizational reforms to enable the effective use of IT, and reforms to social systems and industrial regulations to encompass new technological developments accelerated the shift away from what Freeman calls the old techno-economic paradigm towards a new one. The increased growth rate of labor productivity and the substantial boost in capital productivity seen in the IT industry represented quantum leaps that were to spread throughout the economy. Meanwhile, links in science, technology and R&D were encouraged, creating innovation chain reactions, which would bring about qualitative changes in the greater economic system and social system.

In the late 1990s, the IT industry and related sectors were the driving force behind the growth of the US economy. In 2000, however the inflated share prices that had underpinned that growth went into a freefall, and the US economy, that had until then been reliant on what were now recognized as over-valued high-tech share prices, looked set to enter a period of low growth. Yet there is no denying that the technological innovations that will have a significant impact on the economy in the decades to come are those knowledge-intensive innovations in IT, biotechnology and new materials. Research and development in these fields has become vital, with increasing fusion and integration between technologies in all industry sectors. And advanced Internet-based networks that link these same industry sectors will form the infrastructure critical to the growth of knowledge-intensive industry. Accordingly, we must envisage a framework in which we can use this paradigm as a foundation for environmentally-friendly sustainable development.

### 1.3.2 The Digital Economy and Creation of e-Marketplaces

The core of Internet business at present is business-to-business, or B2B as it is known. The scale of the B2B market is vastly different to that of the B2C (business to consumer) market. Upon closer inspection, electronics and IT-related products, or electronic component purchasing, have by far the largest market share. This is followed by auto parts procurement and construction. Initially B2B begins between specific companies with orders placed over the Internet with existing suppliers. Eventually, more and more parts suppliers from all around the world will be included and companies will be able to produce better products at low cost and procure parts by selecting companies that always deliver on time.

Thus, e-Marketplaces have a vital role to play. An e-Marketplace is an environment that can be accessed by companies anywhere, enabling orders to be placed and received over the Web. e-Marketplaces for automotive parts and electronic components are already in place around the world.

e-Marketplaces generally have four functions. The fundamental feature of current e-Marketplaces is (1) the use of an "aggregator" or "catalog" for transactions with fixed prices that is suitable for handling the sale of standard products between multiple buyers and sellers. While these transactions do indeed have a large share of the market, this function alone provides little incentive for new capital investment. Therefore, e-Marketplaces should also feature the following three functions along with the first: (2) seller-driven "auctions" for the trading of rare or used goods; (3) buyer-driven "reverse auctions" where multiple sellers compete to sell products; and (4) exchanges where custom-made goods with standard product specifications are traded under mutual conditions. In the future these four functions will be integrated into Web solutions that will enable one-stop processing via a browser. Integrating these functions is sure to make e-Marketplaces truly user friendly for both buyer and seller.

At present, the primary function of e-Marketplaces is the placing and receiving of orders, but in the future they will need to be linked to payment settlement networks. They will also need to provide insurance options, especially insurance against damages or loss, and credit functions. Currently major companies mostly place orders with small and medium enterprises (SMEs) that receive those orders and supply the parts. Insufficient capital and poor capital liquidity are a potential problem for these SMEs. A bridging finance function will need to be incorporated. Currently, trading companies, non-banks and regional financial institutions provide bridging finance to SMEs, but this function is not available in e-Marketplaces. More and more e-Marketplaces now offering credit functions and distribution links are also important. If orders are processed over the Net, but distribution is not efficient, then this will greatly impair overall efficiency.

IT investment is essential for providing these functions, even if they are outsourced. Existing organizational structures need to be reviewed and business practices and industrial relations (labor-management relations) should be overhauled to stimulate this investment.

The IT revolution has not only had implications in the economic arena by changing company structures and behavior patterns, it has also affected people's lifestyles. Economic systems are going through sweeping changes on a global scale as they move toward a new economic order.

The full-scale implementation of B2C will take some time given the many system-related issues that need to be resolved on an international level, such as the systemization of electronic authentication and digital signatures, and the protection of personal information. But if we succeed in coordinating the systems of individual companies, then B2C will really take off. B2C is already making impressive inroads in certain sectors, including the automotive industry, computers, peripheral equipment and travel, and B2C is expected to post significant growth in the future in consulting, banking, insurance, securities, education, health care and other service sectors.

The nature of e-commerce generally means that the parties do not interact face-to-face, and as such, establishing trust in the market has become a major issue. This issue is even more crucial in B2C transactions, and the formation of Web-based communities that facilitate horizontal communication has come into focus. For example, by including functions for consumers to give their opinions and exchange product or service evaluations, and by having the service provider respond promptly to any questions or reservations, businesses can boost the credibility of Web-based business, and at the same time bolster the credibility of businesses that advertise on the Web, and provide useful market information. This will also work to improve investor evaluations. Reliable third-party checks must be incorporated to provide such a place for doing business, which means that elements external to the market will have to be introduced.

Now that we have described the international climate for the digital economy, we next look at the significance of information networks in sustainable development based on global environmental protection, the major issue of the twenty-first century, and exmine information networks in relation to industrial organizations and the governance of local communities. As explained by von Weizsacker (1992), information technology can be applied to environmental issues. In the first place, it can be used to accurately monitor and analyze ongoing compound environmental degradation. Also, information technology can make it possible to tailor production to adapt to fluctuations in demand. It can also facilitate advanced quality and inventory control, which in turn can lead to the more efficient utilization of energy and resources. In other words, information technology can improve energy productivity (gross domestic product per kilojoule). Moreover, systematic upgrades of urban transport facilities, telecommuting (working from home) and videoconferencing can all work to reduce the volume of traffic, attenuating air pollution. Yet it must be noted that there is no guarantee that the growth of the digital economy, which grew out of the IT boom, will be consistent with sustainable development. Therefore, the next section will focus on how we should integrate the digital economy and sustainable development, and provide a vision for future society.

## 1.4 Creating a Sustainable Society

Global action to deal with environment problems will be essential in the twenty-first century. Today's global environmental problems call for a reexamination of our modern-day materialistic civilization, and ensuing global environmental protection will require the adoption of new socioeconomic systems and new ways of life. As such, this chapter examines the relationship between the digital economy and e-Democracy and global environmental issues, and exmines how information networks can contribute to a new form of social development that realizes environmental protection.

### 1.4.1 Zero Emissions and the Formation of New Industry Relations

In 1994, the United Nations University (UNU) announced the UNU Zero Emissions Research Initiative (Anon. 1994). Under this initiative, "zero emissions" is defined as "the complete elimination of all emissions into the hydrosphere and atmosphere, and the conversion of waste products from one industrial sector into reusable materials for another sector." The initiative states that to survive the twenty-first century, industry must redesign manufacturing processes, give priority to the use of recyclable materials, and ultimately aim to reduce emissions to zero (Anon. 1994: 2).

In other words, industry must subject the types of materials it uses, and their manufacturing and distribution processes to a strict reexamination, and move toward a zero emissions-based industrial structure in which no waste products are generated at any stage of the production and distribution process. This will require industry to undergo the massive task of reviewing conventional production and distribution processes and creating new industry relations.

For now, the United Nations University's approach is to encourage step-by-step efforts in local communities and then to link these efforts together. One example given by the United Nations University is the use of sugar-based cleansing agents in the brewing industry (Anon. 1994).

The sugar market is currently suffering from a situation of oversupply, as growth in the demand for sugar as a sweetener has fallen in developed countries. Yet many developing nations have no choice but to pursue monoculture production for sugar. Poter and Brown (1996) have examined in detail the interactive relationship between the international economy and environmental problems that impact developing nations (Poter and Brown 1996). According to their observations, with the exception of newly industrializing economies (NIES) that are successfully moving into industrialization and resource-exporting countries that are blessed with underground oil reserves, the majority of developing nations suffer foreign currency shortages. Capital injections from developed countries fell dramatically in the 1980s, from $41 billion in 1981 to $32.4 billion in 1988. In addition, the high-interest policies of the United States increased installment payments, which made it difficult for many developing nations to fulfill their payment obligations. The falling prices of primary produce throughout the 1980s was also a major problem for many developing countries, with most earning, over 75% of their revenue from this source. This typifies the international market for agricultural produce. The sugar market represents another typical example in which sugar prices have fallen below cost due to oversupply. If sugar-producing countries do not export this sugar, even at below-cost prices, they can not obtain the dollars they need to settle international currency transactions, and so these countries are left with little choice but to encourage sugar exports.

However, developments in biotechnology have unearthed new possibilities for sugar. Alkyl polyglucose (AGP), which is found in sugar derivatives has exceptional cleansing properties, with a cleansing power roughly equivalent to those agents produced from petrochemicals or from pollution-producing monocultures (coconut plantations). However, sugar-based cleansing agents decompose faster than conventional cleansing agents, with 99.7% decomposition within 48 hours. As such, they place relatively little strain on the environment, and can help to curb river pollution. In addition to this application, sugar can also be used in pharmaceuticals products and plastic materials, and sugar demand is now expected to see significant improvement. With appropriate capital investment, sugar industries in developing countries can look forward to new opportunities for growth (See Fig. 1).

The link between the cleansing properties of sugar and the brewing industry is indeed very interesting. Beer is currently produced all around the world, and is a process guilty of polluting the environment. The beer kegs and brewing kegs need to be washed out with powerful cleansing agents to ensure that they meet stringent hygiene standards. Naturally, these powerful chemical agents are potentially harmful to consumers which means the kegs need to be cleaned a second time, resulting in large quantities of water being used to rince away the chemical agents.

**Fig. 1.** New uses for sugar

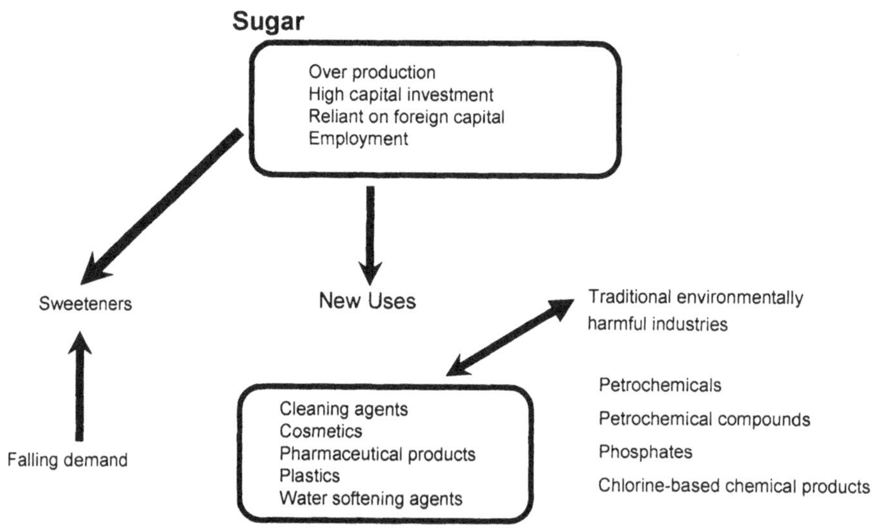


To date, sugar has mainly been produced as a sweetener, but demand for sugar sweeteners has been falling for some time. Therefore, new uses for sugar must be found to aid sugar-producing nations and sugar producers.
In recent years, however, developments in biotechnology have revealed that sugar can be used to produce biodegradable cleansing agents, cosmetics, pharmaceutical products, plastic materials and water hardening agents. What's more, this production requires only a relatively modest investment in plant and equipment. This discovery could be used to facilitate value-added production in developing countries and imposes an insignificant environmental burden on the exporting countries.

Source: The United Nations University, Zero Emissions Research Initiative, 1994, p.18

In contrast, the use of sugar-based cleaning agents simplifies this cleaning process. Sugar-based cleaners have sugar content, and the residue left after beer brewing processes is mostly high-grade protein. This sugar content and residue can be used for livestock breeding and fish farming. However biotechnology has not yet become sufficiently widespread to make these new sugar applications a reality. The introduction of new chemical processing technology could integrate sugar systems with brewing and farming systems. This will then bring the sugar, brewing and farming industries together, thereby creating all-new industry relations (See Fig. 2).

**Fig. 2.** Beer Manufacturing and Farming Systems

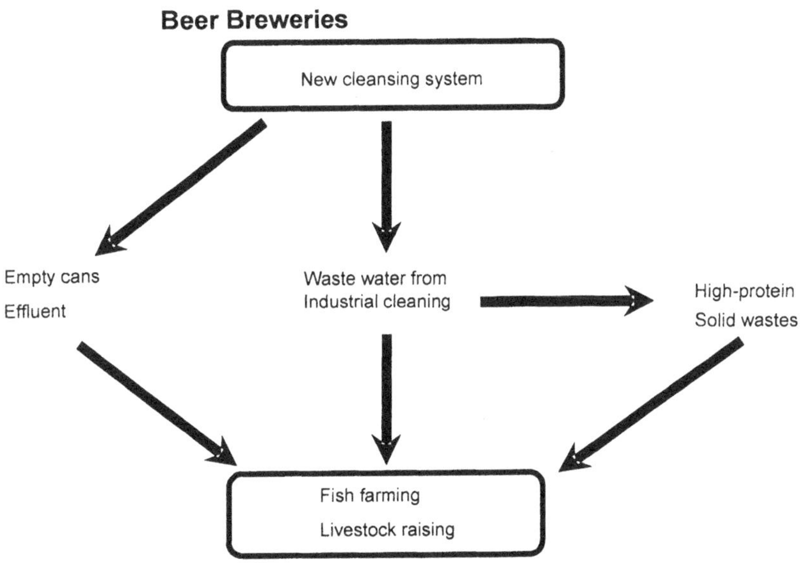


When sugar-based cleansing agent to clean beer brewing kegs and empty cans, the discharge liquid includes sugar content from the cleaning agentand high-grade protein solid waste from beer residue. This waste discharge can be transported to nearby fish farms or cattle farms and used effectively as feed. If aquaculture facilities are located adjacent to beer manufacturing facilities, then the waste from sugar-based cleansing agents and solid protein residue can be directly administered as feed. The feces of fish fed on this feed can fertilize marine plants.

Source: The Nations University, Zero Emissions Research Initiative,1994, p.13

According to the United Nations University, the government sector (industry policy makers and local governments) will benefit most from the zero emissions initiative mentioned earlier, and then ultimately so will the taxpayers. Calculations by the UNU show that, although to date the government has invested heavily in creating industrial estates, if industrial infrastructure were to be upgraded under zero emissions policies then companies would use only around one tenth of the volume of drinking water they used previously. Concentrated effluent processing facilities would no longer be needed as effluent would be able to be recycled and reused within the company, and the less energy-intensive production processes would reduce power consumption to one fifth of the previous level. Estimates suggest that this could reduce the cost of building and maintaining industrial estates to 20% of current levels. That would make it an extremely significant industrial policy for the economic growth of developing nations facing currency shortages.

Next we exmine how government will change using information technology and information networks, and what sorts of changes will accompany this in local communities. These observations are then used to comment on sustainable regional development.

### 1.4.2 The e-Municipality Concept and Regional Development

In many counties, there is a vigorous push towards the establishment of electronic government (e-Government) and electronic local government (e-Local Government). In Japan too, the e-Government and e-Local Government concept is a crucial element of their e-Japan strategy. The aim and indeed the benefit to society of e-Government and e-Municipalities is the creation of effective demand in the IT sector. However, the benefits do not stop there. IT can be effectively utilized to overcome bureaucracy in the government sector or the public sector, build a more efficient administrative organization, and to improve the quality of public services. IT and networks must also be used effectively to promote local democracy. We need to establish a citizen-centered government that is founded on IT and networks.

However, measures to compensate for the lack of financial and human resources are vital for the implementation of e-Local Government. Figure 3 depicts the type of e-Municipality system that would be established in an e-Municipality in Japan. Financial accounting, taxation, human resources and other key systems are already operated independently by local governments. Also, all employees use terminals connected to intradepartmental local area networks (LANs), and many local governments have finished setting up groupware and staff portals. But what needs to be established now, as shown in the central part of the Fig.3, is a portal that will enable external electronic applications and electronic purchasing, as well as facilities for connection to document management and electronic approval systems, electronic purchasing systems, electronic application processing systems and multi-payment systems for handling taxation payments. Also needed are local government wide area network (LGWAN) connections to link all local governments, and authentication systems. The cost of establishing and operating these systems through each independent local government will be exorbitant and the effort will involve vast numbers of personnel with a sophisticated knowledge of information systems. The idea is therefore for local governments to work together with companies to introduce these systems, and then to operate them jointly; in other words, the system for joint outsoursing to a government application service provider (ASP) joint outsourcing. As is evident in Fig. 4, there are a number of different modes of outsourcing and the most suitable mode for the particular region and local government conditions must be selected. For example, there may be cases in which integrated systems are constructed for the government-designated cities (12 major cities) and local governments in those surrounding areas. Or prefecture governments may work together with all their municipalities to develop a single system.

Figure 5 depicts electronic local communities that are based on data centers and the networks connected to these data centers. First, data centers will be built to provide a secure environment for server management, where ASPs will be concentrated. Then the government will pay the service charges for the provision of server management, application software and to ensure security. This will save on the high initial investment costs, and will mean that ASPs will provide services for software upgrades and the like.

Meanwhile, many SMEs in the now very advanced digital economy are visibly active in the B2B sector, and are hoping to upgrade to the B2C sector as well. The establishment and management of information systems involves the provision of secure access via authentication and digital signatures, the installation and operation of sophisticated software, and storage (saving of transaction records and documents). All this adds up to a very serious hurdle for SMEs, with their very limited capital and human resources. Yet if data centers and ASPs are effectively used to manage the servers of local companies, the linchpin of both the local economy and the local community, then this will facilitate their integration into the digital economy.

Another issue is the growing prevalence of Internet crackers (people who use software or network glitches to cause damage by interrupting and disabling

**Fig. 3.** E-Municipality Framework

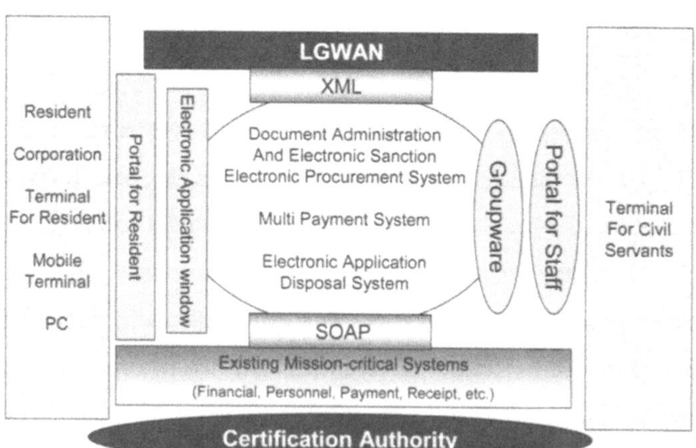


Local residents will be able to use third-generation and fourth-generation mobile phones and PDAs (Personal Digital Assistances) etc., with powerful data transmission capabilities, touch panel terminals installed by local governments (systems based on the same principles as station vending machines and ATMs, etc.), and personal computers to lodge electronic applications, and apply for subsidies and health care services. Companies, too, will be able to use the same sorts of networks to make applications and conduct business transactions with government offices.

computer systems). With the proliferation of computer viruses, it is now extremely difficult, not to mention costly, for members of the general public to manage their website servers themselves. That's were data centers enter the picture. The servers can be stored at the data center where the website of each establishment or household is managed, guaranteeing security. This will then mean that citizens will be able to access public information, in addition to being able to manage their own Web content and post personal information on their own sites. Let's take for example a case in which a family member is admitted to a health-care facility. Those at home can communicate with the facility and when the patient is discharged the patient too can remain in contact to the hospital. This kind of system can also form the basis of community activities networks.

**Fig. 4.** Modes of Joint Outsourcing

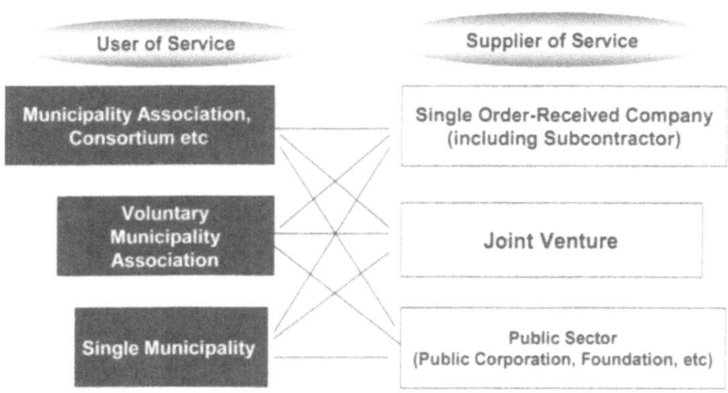


The system configuration shown in Fig. 3 will require significant plant and equipment investment if it is to be implemented by each individual municipal government. Moreover, maintenance and management of this system and provision of services will require sophisticated data processing technology, network management skills and security function capabilities. It therefore needs to be jointly implemented through collaborative efforts between municipal governments with standardized administrative processing procedures. It will also be necessary to outsource daily operations to private-sector companies with highly skilled IT technicians. Fig. 4 shows the outsourcing modes.

Reference: Research council on guidelines for outsourcing in public-sector IT (Ministry of Public Management, Home Affairs, Posts and Telecommunications)

The ultimate aim of the e-Municipality concept is the development of local democracy. Industries, government, local residents, and research and educational institutions need to develop a cooperative system for the management of the local community. In doing so, they will develop a new local community through the exchange of ideas between people of different stations and the encouragement and utilization of individual creativity. All citizens need to be able to participate in this system without impediment. We must build a universal community in which the elderly and physically disabled too are free to voice their opinions and contribute to the policy implementation process. We need today a design model for such a system. Put simply, we need to establish electronic local communities in which the major local industry, government, citizen and academic groups collaborate.

**Fig. 5.** Creation of Electronic Local Communities

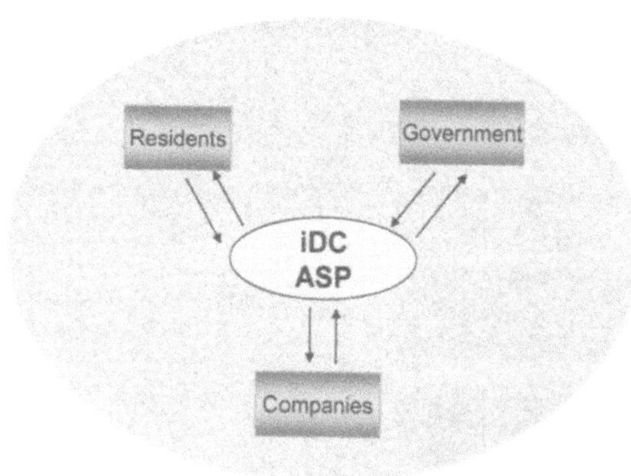


e-Municipality administrative information systems will be used jointly and therefore the facilities delivered by these systems must be subject to stringent security management. International Organization for Standardization (ISO) standards require earthquake-proof structures (quake-absorbing structures), two power supply systems, private power generation equipment, biometrically controlled access, and the latest surveillance systems. Facilities that meet these criteria are called Internet Data Centers (iDCs). iDCs provide and manage not only government administrative systems, but also corporate information systems and servers for the general public under stringent control conditions. Fig. 5 depicts a local electronic community with iDCs providing nodal functions.

In such a system, the role of the coordinator is critical. Local governments need to be more involved than ever before, and NPOs are also likely to play an important role. As is evident at Mitaka City in Tokyo, Ichikawa City in Chiba Prefecture (Japan), and at San Jose and Palo Alto in California State (United States), NPOs have formed new cooperative relationships designed to put pressure on governments, and their efforts to expand local community management are gathering force. In the future, there is great potential for the new public sector that features collaborative relationships between citizens, government and industry to take over many of the tasks previously handled by the government or local administrative authorities. Many local governments are finding it very difficult to win financial backing for new work with the progressive aging of Japanese society.

NPOs exist somewhere between governments, local authorities, private enterprise and citizens, and as the node of local communities, they are likely to become a key entity with the critical role of using networks effectively for the purpose of policy assessment, and even policy implementation. The term "e-Democracy" means citizen participation in the process of policy formulation, policy making and execution through the use of IT. E-Democracy calls for not only systemic upgrades, but also for citizens themselves to become more informed through e-learning (Web-based education using the Internet). In this regard, the electronic local communities described earlier will form the foundation for the evolution of democracy as we know it. As such, the next part of this chapter will look at the social functions of NPOs.

### 1.4.3 A Knowledge Society Based on NPOs and Networks

The digital economy is a seamless and global entity, which differs markedly from the more familiar international economy, where individual nations that make up the international economy each support their own particular systems. Accordingly, in principle, international frameworks and agreements in the digital economy should be driven by the private sector, and as such NGO activities will be very important. NGOs are characterized by their function as organizations that do not operate in either the public sector or the private sector. Instead they incorporate elements of both sectors, or in other words, are a hybrid of the two sectors, and fall somewhere between the two. (See Fig, 6)

NPOs that function as community network nodes will play a leading role that differs both from that of the government and that of corporate enterprises primarily concerned with commercial gain. They will have an increasingly crucial function in the digital economy. We need to first clarify the role of intermediary entities like NPOs before designing a system that will form the foundation of the digital economy.

**Fig. 6.** Societal Role of NPOs

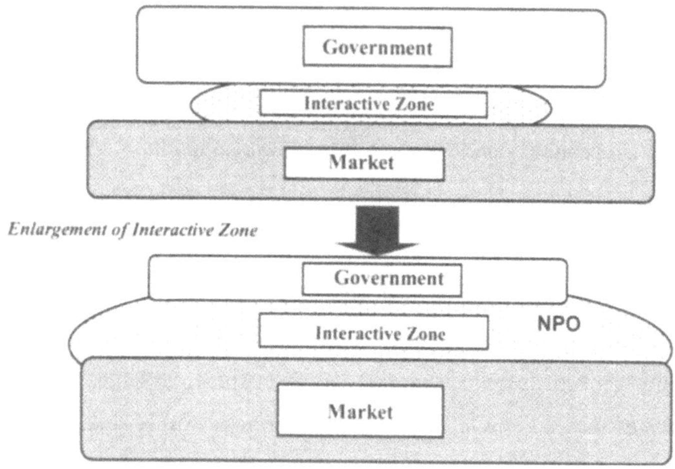


The upper portion of this diagram shows that the government-driven public sector and the market have hitherto played a large role in the social system. However, even in the conventional social system, there has always been an intermediary coordinating entity between the market and the government sector. In America, lobbyists have traditionally filled this role.
The lower portion of the diagram shows the shrinking role of the government sector and the growing role of the market, and the associated enlargement of the intermediary coordinating entities. These coordinating entities will be restructured to be more democratic, and NPOs must fulfill a social function by playing an active role in the open coordination process through information disclosure.

In more specific terms, NPOs basically have three main functions. Their first function closely resembles that of commercial activities. In this case, NPOs will have a coordinating role as community network nodes. They facilitate the establishment of a rough set of guidelines based on the mutual evaluations of diverse entities, and coordinate their various interests. As exemplified by the success of the Silicon Valley model in the United States, having NPOs in a coordinating role will promote the development of technological innovation and business models in private enterprises. In Finland, the leader of the IT revolution in Europe, intermediary organizations formed through collaborative relationships between industry, government, academic and citizens, known as "forums," play an instrumental role in coordinating knowledge in these four sectors. These forums are non-profit organizations, and in some ways fulfill the same role as the coordinating NPOs in Silicon Valley.

The second function of NPOs involves public activities. In this case, NPOs evaluate activities and take on a navigating role as a TTP, or Trusted Third Party. As TTPs, NPOs act as citizen participation-based intermediary evaluation entities and complaint handling entities who apply semi-public regulations for monitoring and handling disputes, as do the Better Business Bureau (BBB) in the United States and ombudsmen in Northern European countries. They can play an important part in privacy protection and consumer protection.

The third function of NPOs involves making social valuations and building reliability through mutual evaluation amongst network members. In this way NPOs can be classified as entities that create social value. NPOs that create social value can promote the activities of global networks and advance open-source technology innovations. This is exemplified by their involvement in the evolution process of LINUX, one of the open-source operating systems. Their activities may also make viable the concept of a social system that targets environmentally-friendly sustainable development.

The digital economic climate highlights the necessity of hybrid NPOs that use networks to facilitate relationships and interaction between NPOs. If NPOs that make social evaluations can be effectively incorporated as coordinators, technical standards for environmental protection may be produced. Also, having NPOs take on both a coordinating role and a TTP role will stimulate efforts to integrate local welfare activities and environmental protection activities that make effective use of IT and networks.

Moreover, these NPO coordination and valuation activities will strengthen the relationships of trust between local residents and companies on the one hand and local authorities on the other, allowing them to exchange information and knowledge and develop policies together for regional social development. This in turn has the potential to dramatically change the economies, cultures, welfare policies and environmental policies of local communities.

Today there are many local currencies circulating through local areas around the world. While these local currencies generally are not redeemable for real currency, they can be used in exchange for a variety of services, goods, knowledge and information. They are also used to assess the value of environmental protection activities, welfare activities, and other activities in local communities. If electronic local currencies that are not redeemable for real currency are introduced by NPOs for environmental protection, then using this electronic local currency as a medium of exchange will mean that environmental protection activities can be valued in the local community and people who acquire that currency can exchange it for daily conveniences they require.

This is a kind of localized reciprocal system for environmental protection. But let's assume that environmental NPOs issue a local electronic currency over the Internet, one that is not redeemable for real currency, and that this currency comes into widespread use. This localized reciprocal system then has the potential to spread globally, and from that point on the electronic local currency will no longer be localized. This local currency can be used as a means of payment for the global exchange of software, databases, electronic publications and more. If a significant number of people around the world participate in this system, and one electronic local currency becomes "a more saleable good" (Carl Menger) than the others, this will provide widespread incentive amongst people to hold a certain proportion of their assets in this electronic local currency.

If environmental NPOs expand circulation of the electronic local currencies they issue, there are a number of points that should be heeded to ensure that these currencies function correctly as a means of exchange[1].

---

[1] F. A. Hayek, in his 1976 work "The Denationalization of Money," proposes that national ties on money be lifted, that the government monopoly on currency circulation be broken down, and that the private sector be permitted to freely circulate money. The most important aim of his proposal was to limit credit creation, and to prevent monetary chaos, such as inflation that stems from economic activity.

### 1.4.4 Knowledge Networks and Environmental Protection

The Zero Emissions Initiative involves forming industry relationships in which input equals output, so as to create an environmentally-friendly industrial structure that does not generate any waste products. In creating an industrial structure free of waste products, one of the most important issues is the organization of corporate groups that are involved in different industries. Production management information and waste management information channels are imperative.

Information technology can process large volumes of diverse data into readily comprehensible formats, and can be used as a powerful tool to turn information into knowledge. Today information technology is becoming more heavily focused on autonomous distributed networks and interactive communications.

Networks can basically be classified into two types: networks for efficient data flow management, and networks for non-data-based knowledge formation and communications.

These two types work together and tend to merge into one. Networks of the latter variety designed for knowledge formation and communication are highly dependent on networks of the former variety to manage and standardize the flow

---

According to Hayek, the currency issuer decides the currency denomination, and must announce the exchange rate of that currency against other currencies. The currency is subject to open competition, and in the face of fierce competition, the issuer must minimize fluctuations in the value of the currency they have issued to the greatest extent possible to ensure its survival. If the value of the currency fluctuates wildly, then it will not be able to fulfill one its basic functions as money—namely, to act as a store of value—and will fail to win the confidence of the public. Therefore the bank of issue is left with little choice other than to cautiously create credit, and this may cause inflation.
Moreover, there is also the possibility of credit creation through deposit creation by non-issuing banks, but Hayek believes that if issuing banks take steps to disallow the conversion of new deposit currencies created by non-issuing banks and other currencies, then the legal reserve ratio of non-issuing banks will be forced up to almost 100%, rendering any credit creation virtually impossible. This, he claimed, could restrain monetary chaos caused by inflation and the like.
In the event that environmental NPOs freely issue electronic forms of local currency and if there is open competition between the issuers, Hayek's theory on the denationalization of money would be worth looking at when considering autonomous value assessments made in societies targeting sustainable growth. In other words, the reserve ratio of electronic local currency needs to remain at 100% to prevent credit creation. Otherwise, that currency will quickly lose its credibility.

of that information. Meanwhile networks of the former variety that were originally designed for data processing provide the foundation for knowledge formation. Through this interaction, data management networks and knowledge formation networks merge and have the potential to encourage the formation of new value judgments and societal norms. NPOs have a significant role to play in the formation of these societal values.

As has been recounted, the growth of e-Marketplaces has sparked the formation of a growing number of global networks. The formation of these kinds of networks means that companies are sharing their information and knowledge, including production management information, purchasing management information, waste management information and financial management information with each other, and are thereby facilitating the formation of new knowledge.

What is waste for one company in a particular industry may be able to be used as a production material or ingredient for a company in another industry—one man's trash might be another man's treasure. If governments, local authorities and international entities can encourage appropriate investment through discriminative tariffs, subsidies and finance, then this will make it relatively easy for companies to use e-Marketplaces to form networks with companies from other industries, and this is what they need to do to accomplish their zero emissions targets. In other words, global company networks may move towards industry alliances that are in keeping with the principles of sustainable development. The participation of SMEs in particular in these alliances is of great importance, and as was touched on in the earlier discussion of e-Municipalities, the introduction of data centers and ASPs through local industry and government collaboration will enable SMEs to participate in e-Marketplaces. If appropriate investment can then be encouraged through discriminative tariffs, subsidies and finance, SMEs can be incorporated into the new zero-emissions industry structure. Then we will see a new form of social development that targets both commercial activities and employment, and environment issues, based on the growth of local electronic communities in the digital economy.

The report released by the World Commission on Environment and Development (1987) (commonly referred to as the Brundtland Commission) stresses that dramatically turning around society's values and goals, providing different incentives, and changing decision-making processes will pave the way towards sustainable development that encompasses both economic growth and environmental preservation. The report also indicates that to achieve sustainable development, civic initiatives need to be increased, power needs to be given to civic organizations, and local democracy has to be strengthened. The report also emphasized the necessity of government systems that guarantee civic involvement in decision-making processes.

Given today's complex international climate, what sort of predictions can be made for the future of the global environment and market economy? Inter-company networks are being formed within certain economic zones; companies, financial systems and public sectors are being integrated in local areas with NPOs positioned as the nodes or junction points; and there are moves to establish and strengthen cooperative relationships among companies, universities, local authorities and the community. If more concrete initiatives emerge for civic organizations in local areas and agreements on environmental protection become established, the environmentally friendly growth of local communities can be envisaged, as highlighted by the Zero Emissions Initiative.

Local community-centered environmental activities, naturally, will be inextricably linked to global networks that link local activities. When visualizing and planning environmentally friendly development in local communities we must recognize that it will be based on global interactive relationships. A system must be put in place in which local residents, companies, local authorities, and the government cooperate under the banner of participatory local democracy, to work toward achieving environmentally-friendly development in local communities, and in which each local community uses the Internet to form global networks, and exchange information and share knowledge.

For example, perhaps local residents should have corporate status like NPOs, and establish organizations through civic sector initiatives with a clearly defined locus of responsibility. These organizations could then be used to monitor and evaluate environmental protection in the community. Using LGWAN, governments will be able to have common access to all of the statistical data and laws and ordinances held by the government or any local authority. This information will also include a range of information that relates to the environment. Then this information will be stored in a server for public information so that it can be accessed as a database at any time by local administrative entities, residents and companies. Moreover, NPOs will monitor environmental activities in local areas through different entities, and will make evaluations based on set evaluation criteria. So these NPOs act as trusted third parties (TTPs). Governments can use these third party evaluations to formulate assistance packages, such as subsidization for environmental activities. Local residents, companies, local governments and research and educational institutions will utilize these NPOs as community nodes (junction points), and NPOs must interact and coordinate knowledge with government policies, and incorporate e-learning techniques into education initiatives to increase public awareness of the importance of sustainable development. E-learning will enable people to access environmental information and knowledge relating to environmental protection anytime and anywhere. Forums for debate such as electronic bulletin boards should also be provided, and information acquired through such means can then be shared and used in environmental activities. There are great expectations that NPOs will also lead this charge, which will mean that NPOs will also come to be responsible for creating social value.

In this way, we need to actively use information systems to establish a multi-tiered mechanism for environmental protection, and then incorporate NPOs into global strategies for sustainable growth in which nongovernmental organizations (NGOs) act as international nodes. NPOs formed through collaboration between local residents, companies, and local authorities will evaluate environmental activities, but then multiple action groups should be established through these civic initiatives, and these groups should be incorporated into global networks. While these groups will be locally based, we need to establish a framework that promotes both economic development and environmental preservation through the global partnerships forged by local areas, and the zero emissions initiative provides a prime example of this.

Think back to Piore and Sabel's theories (1984) on the local industry networks seen throughout nineteenth century Europe, which were detailed in Section 1. There we outlined the preconditions for a network-based industry structure with the coordinating functions summarized in three points. Evidently, sustainable regional development based on global networks now has increasing potential to replace mass production, mass consumption and mass disposal.

## 1.5 Conclusion

Many environmentalists have repeatedly emphasized that the Internet is becoming a vital resource for environmental activities. E-mail has already become an indispensable tool for those dealing with environmental and social issues and NGOs around the world are using information networks to access environmental information, debate environmental issues, coordinate activities and swap news. As such, the Internet, the world's largest network, has become the major resource for people working to preserve our environment. In addition, the digital economy can facilitate the formation of new industry relations that will form the foundation of the UNU's Zero Emissions Initiative, and like e-Marketplaces, the digital economy continues to expand on a global scale.

With the expansion of these kinds of information networks, the various networks of people tackling sustainable development are mutually influential and are gradually merging to form a massive global electronic community that completely transcends the barriers posed by distance. As the threat of global environmental degradation becomes ever more serious, in the long term, these global networks may wield significant power to be able to link regions to the world and to foster new societal norms that overcome nationalistic interests. Citizens, governments, local authorities and companies must cooperate under the banner of participatory democracy to work towards the environmentally-friendly development of local communities, and these local communities needed to be linked to form a global network for environmentally-friendly social development. It may be a little idealistic, but if developed countries promote financial aid and technology transfers to developing countries, while at the same time local communities recognize each others independence and diversity and join together to form powerful global networks, then surely we can break free of the tangled web of nationalistic interests that plagues the world today, and develop a new set of common societal values that will take us into the future.

## References

Accenture (ed.) (2004) e-Government Leadership: High Performance, Maximum Value. Accenture

Anon. (1994) Zero Emissions Research Initiative. United Nations University, Tokyo

Anon. (2002) E-Commerce and Development Report 2002. United Nations Conference on Trade and Development, Paris

Brynjolfsson, E and Kahin, B (2002) Understanding the Digital Economy. MIT Press, Cambridge, MA

Cooper WW, S. Gibson T.D. and Phillips F (1997), Impact. Quorum, Westport

Freeman C (1992) The Economics of Hope. Pinter Publishers, London

Hayek FA (1978) Denationalisation of Money. 2nd ed. The Institute of Economic Affairs, London

Holmes D (2002) e-Gov: e-Business Strategies for Government. Nicholas Brealey Publishing, London

Lomborg B (2001) The Skeptical Environmentalist: Measuring the Real State of World. Cambridge University Press, Cambridge

Matsushita K (2002) Environmental Governance: The role of citizens, companies, local government and the central government (in Japanese). Iwanami Shoten, Tokyo

Menger, Carl. (1871). Grundsatze der Volkswirtschaftslehre (in Germany). In: Gesammelte Werke, Bd.1, Tubingen, 1968

Morigaki T, Matuura K and Sudoh O (2004) An Analysis of Detailed Electronic Time-Stamping Using Digital TV. In: IEEE International Conference on e-Technology, e-Commerce, and e-Service (EEE04), pp 277–284

Organisation for Economic Co-Operation and Development (OECD) (2002) Measuring the Information Economy 2002. OECD, Paris

Pajarinen M, Rouvinen P and Yla-Anttila P (1998) Small Country Strategies in Global Competition: Benchmarking the Finnish Case. ETLA and SITRA, Helsinki

Piore MJ and Sabel CF (1984) The Second Industrial Divide: Possibilities for Prosperity. Perseus Book Group, New York

Poter G and Brown JW (1996) Global Environmental Politics. 2nd ed. Harpercollins, New York

Shinkai I, Kasuga M, Yamada E, and Kanaya T (2002) The E-Democracy and Regional Development using ICT (in Japanese). Shogakkan, Tokyo

Sudoh O (1995) The Interactive Network Society: The information technology and social Evolution (in Japanese). Yuuhikaku, Tokyo

Sudoh O, and Gotoh R (1998) Electronic Money (in Japanese). Chikuma-Shobou, Tokyo

Sudoh O (2001a) Digital Economy and Sustainable Development. In: Anglo-Japanese Academy (ed) Conference presentation papers on national, regional and global transition: a common agenda for Anglo-Japanese relations in the twenty-first century. University of Sheffield and University of Tokyo, Tokyo and Sheffield, pp 265–271

Sudoh O (2001b) The Digital Revolution and the Evolution of e-Government, In: Institute for International Policy Studies (ed.) The International Symposium Presentation Papers on The IT Revolution: Challenges from innovation in information and communication technology and the role of government.Institute for International Policy Studies, Tokyo, pp 1–17

Sudoh O (2002) Wealth of Networked Nations. Look Japan 558:12–14

Sudoh O (2003) The Digital Economy and New Governance. Review of Media, Information and Society 8:17–39

Sudoh O and Deguchi H (eds.) (2003) The Institutional Design for Digital Network Society (in Japanese). NTT, Tokyo

Sudoh O (2004a) Strategies for Sustainable Development and the Formation of NPO Knowledge Network. Review of Media, Information and Society 9: 11-30

Sudoh O (2004b) The Knowledge Network in the Digital Economy and Sustainable Development, In Sudoh O (ed.) Reviewed Conference Papers on Establishment of the

Digital Economy and Evolution of the Socio-Economic System, The University of Tokyo, Tokyo, pp 27–41

Takayama Y (2001) The Labor Divide (in Japanese). Nihon Keizai Shimbun, Tokyo

Takehiko K, Sumiaki M, and Ueda K (2002) Introduction to Environmental Science (in Japanese). Iwanami Shoten, Tokyo

Ueda K (1996) Environmental Economics (in Japanese). Iwanami Shoten, Tokyo

US Department of Energy (2002) Renewable Energy Annual 2001: With Preliminary Data for 2001. Energy Information Administration, Department of Energy, Washington DC

US Department of Commerce (2002) Digital Economy 2002. US Department of Commerce, Washington DC

Weizsäcker Ernst U. von (1992) Erdpolitik: Ökologische Realpolitik an der Schwelle zum ahrhundert der Umwelt (in Germany). 3. aktualisierte Auflage.Wissenschaftliche Buchgesellschaft, Darmstadt

World Commission on Environment and Development (1987) Our Common Future, Oxford Press, Oxford

# 2 Controllability of Technological Paradigm Shift and Sustainable Growth: An Empirical Research on 1970–2002 Japanese Economy

Shungo Sakaki

School of Media Science, Tokyo University of Technology, 1404-1 Katakura Hachioji, Tokyo 192-0982, Japan, sakaki@media.teu.ac.jp

**Summary**

By computer simulation applying a social learning dynamics model, it is formerly argued that depreciation rate management could maintain sustainable growth with technological change over the long term. The primary objective of this chapter is to argue the results of innovation management in the Japanese economy. Several results could be suggested by an empirical examination as follows. Japan's asset securities reports demonstrate that Japan's listed manufacturing companies have satisfied the conditions for managing sustainable growth. However, the private sector capital stock statistics demonstrate that Japan's unlisted manufacturing companies might have failed to manage sustainable·economic growth with technological change by depreciation rate since the late 1990s. It can be pointed out that one contributing factor to the IT productivity paradox is inefficiency caused by systemic inconsistency between Japan's listed and unlisted manufacturing companies.

**Key Words** Technological change, Sustainable growth, Technology diversification, Technology standardization, Systemic inconsistency

## 2.1 Introduction

Innovation is born of intangible stocks of information and knowledge through a series of stages in which this information and knowledge is practically applied. The information then takes the form of tangible and intangible capital stock used as a means of production and management, which are in turn used to generate new commercial goods and services. For sustainable economic growth over the long term, the products of technological innovation that determine the overall national economic system must be disseminated throughout the entire national economy. At the same time, we must manage the evolution of technological innovation to limit any "locking into" existing forms of technology or knowledge stock (stagnation of technical innovation) and to trigger all new technological innovation for the next generation. By applying a social learning dynamics model, we argued that depreciation rate management could maintain sustainable growth with technological change over the long term (Sakaki 2004a, 2004b). When we look at the relationship between technological change management and production benchmarks over time, provided that the results generated by the simulation are correct, we see that when marginal productivity of knowledge stock is relatively high, the depreciation rate keeps relatively low. Conversely, when the marginal productivity is relatively low, we see an increase in the depreciation rate. In this way, sustainable growth can be achieved through depreciation rate management. This chapter focuses on Japan's asset securities reports, as well as private sector capital stock statistics to: (1) consider the two elements of technological development paths—standardization and diversification, and (2) make an empirical examination of the success of technology change management in Japan's postwar economy.

## 2.2 Long-Term Economic Growth and the Innovation System

### 2.2.1 The Concept of Resource Allocation

The problem associated with the kind of technological development described in the Introduction is understood to be one of resource allocation. This means that there is a constant issue of whether to focus on extending existing technology or changing over to entirely new technology. This issue is relevant on a both the microeconomic and macroeconomic levels in terms of research and development in companies and science and technology policies for the national economy.

In this chapter, the choice of how to allocate limited resources for research and development activities is examined as either an "efficiency" approach in which resources are concentrated on a particular area of technology, or a "diversification" approach in which efficiency may be somewhat sacrificed so as to secure various sources of technology. This involves comparing the concentration of resources on existing knowledge stock, which is widespread through the society but has a relatively low marginal productivity, with diversifying resources across new knowledge stock that is far less prevalent in society but has extremely high marginal productivity. We examine what proportions of limited resources should be allocated in each of these approaches to achieve higher economic growth over time. In line with Dosi (1990), this chapter categorizes the former approach as "static efficiency resource allocation", and the latter approach as "dynamic efficiency resource allocation."

### 2.2.2 Technological Change and the Sustainable Growth Path

Later sections of this chapter describe technological change that occurs by the alternation of the two types of knowledge stock over time, and likens this alternation to the process of continual generation, diffusion, and changeovers seen with macroeconomic innovation. According to Yoshikawa (2000), it is generally observed that the growth of individual products or industries stagnates with saturation of demand, and this can be depicted by an S-shaped curve when changes in production volume are plotted in a time series. The marginal productivity of capital stock used to produce individual products and services will decline with demand-side saturation. However, with the emergence of new products and industries with high marginal productivity generated from other capital stock, the economy as a whole can once again experience increased growth rates. In other words, behind the S-shaped growth path observed on a macroeconomic level, there exists a process of alternation between increasing returns and diminishing returns over time, which is generated by new independent individual knowledge stocks that are constantly emerging (Fig. 1).

This chapter suggests that a switch between two forms of knowledge stock will generate the scenario in which new knowledge stocks with high marginal productivity are generated over time and then go on to produce increasing returns over time throughout the whole economy. In reality, as is exemplified by the following simulation results, individual knowledge stocks saturate the marketplace and reduce marginal productivity, thereby contributing to a fall in macroeconomic growth. In contrast, shifting to new knowledge stocks will generate increasing returns once again: a mechanism that will form the basis for this knowledge stock alternation model.

**Fig. 1.** The macroeconomic process of increasing and diminishing returns from stock alternation

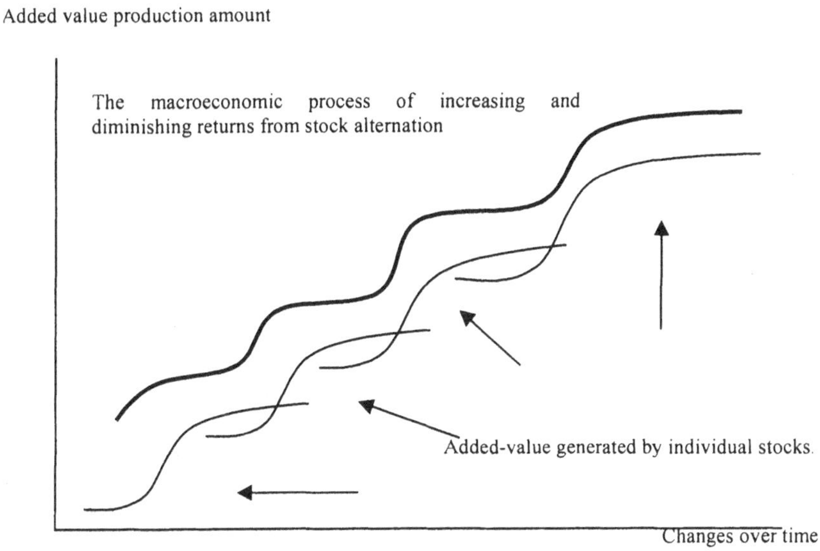


## 2.3 Simulation and Policy Implications

### 2.3.1 Overview of Simulation Model[1]

In the real process of economic development, when new high-productivity technologies emerge, these new technologies tend to replace the older now less-productive technologies, and spread throughout the society to eventually become the technology standard throughout the entire social system. Yet after a time, even these technologies reach maturity and as their productivity declines, new technologies emerge to replace them, and so the process continues.

This chapter suggests that long-term economic growth and the process of technological change, in which new technologies continually emerge and are then

[1] For a formal and detailed description about this simulation model, see Sakaki (2004b).

replaced by even newer technologies, is the result of the alternation over time of two types of knowledge stock, which is the basis of technology. In other words, here we shall try to approximate the continuous state of technological change for sustainable economic growth by explaining a system that selectively switches between the two alternative knowledge stocks A and B over time.

The analysis framework used to solve the problems presented here is based on Deguchi's model of social learning dynamics (SLD) (Deguchi 2000). Following this framework, SLD in this analysis shall involve three interrelated subsystems that indirectly control one another and are used to decide on the fundamental choices associated with the two knowledge stocks.

The first fundamental choice relates to determining the research and development ratio and involves choosing between the alternatives of continuing to use the existing knowledge stock as is, or replacing it with newly generated knowledge stock. In this analysis, it is assumed that a subsystem is made up of groups of agents that choose the knowledge stock renewal alternative, and that the research and development ratio is determined through social learning in this system. The population proportion of the agents that choose different alternatives for knowledge stock renewal will construct an average research and development ratio of all agents belonging to this subsystem. For example, the proportion of agents that will replace all the remaining and existing knowledge stocks with knowledge stock A to all agents belonging to this subsystem is equivalent to the ratio at which knowledge stock A is chosen by an average agent in this subsystem (that determines the research and development ratio) who needs to invest the knowledge stock accumulated to date as source material.

The second fundamental choice is related to agents involved in either production activities or investment activities choosing alternative knowledge stocks based on knowledge stock after upgrades in the first subsystem. This chapter proposes a model in which when selecting from the two knowledge stock alternatives, agents (real/production sector: group X) that select the knowledge stock based on "total output" (total added value) criteria procure funds from agents (financial/investment sector: group Y) that select knowledge stock based on "marginal product" criteria. They (group X) then make plant and equipment investments based on the knowledge stock they have chosen, and engage in production. Agents associated with either of these two sectors are reliant on the population ratio of other agents selecting the same knowledge stock, and then they can make plant and equipment investments based on the upgraded knowledge stocks to generate new gains. However, if they select knowledge stocks in differing fields to each other, then they will be dependent on the population ratio of agents in that field and they will receive gains on a par with those of the previous period after discount or depreciation.

The third fundamental choice involves making a choice on the basis of gross domestic product (GDP), the performance indicator for the entire system, to manage higher growth productivity (growth potential). In this chapter, this involves making a choice based on an assessment of the discount (depreciation rate) on knowledge stock across the whole system. In other words, in this subsystem, the knowledge stock depreciation rate is used as a means of indirectly managing system-wide performance with the macroeconomic index GDP used as a basis for assessment.

Let us consider the two alternatives available to agents in this third subsystem. To do so, we will assume that these agents are able to make the following information processing in their supervisory roles in the greater system that integrates each of the subsystems. By this we mean that through the learning dynamics seen in the second subsystem in which knowledge stock alternatives were chosen, information is obtained that relates to the alternative knowledge stock selection ratio at the start of the term for the real sector and the financial sector. This information is based on the case when gains in the previous period received from selecting different knowledge stock for the real and financial sectors are valued in full at the same level, and the case when they are not valued at all (i.e., when the value of previous-term earnings has depreciated in full, generating a subjective earnings valuation of zero).

Next, we look at the gains when agents in the third subsystem choose from the two alternative cases where the previous-term earnings are valued at their full amount or where they are not valued at all. Then the agents calculate aggregate added value, i.e., the expected GDP for both of these scenarios, respectively, and select an alternative based on this. Therefore, the proportion of agents who select the alternative where the previous-term earnings are not valued at all (or depreciated in full) to all agents belonging to this subsystem can be equivalent on average to a depreciation rate on the knowledge stock.

### 2.3.2 Simulation Results

By simulating the three scenarios where previous-term earnings are (1) valued at their full amount, (2) not valued at all (depreciated in full), and (3) managed using a depreciation rate between the above two extreme scenarios, we will compare GDP, the system-wide performance indicator, over time.

The first scenario represents the case where the effect of the knowledge stock chosen up to the previous period has been extremely strong in both the real and financial sectors, that is, the case where the sunk cost of past investments is exceptionally high, or where past investments are made the most use of strategically or under policy guidance. In this scenario (Fig. 2) as time passes, research and development declines, and as a result both sectors lock into one type of capital stock based on existing knowledge stock. Consequently, GDP surges when this knowledge stock spreads, but eventually stagnates over time as this knowledge stock becomes firmly entrenched throughout the whole system. This leads to falls in productivity, and finally at $t = 8000$ only increases to the high end of the 6% range.

**Fig. 2.** Knowledge stock seletion ratio for the real sector and changes in gross domestic product (GDP) when previous-term earnings are valued in full amount

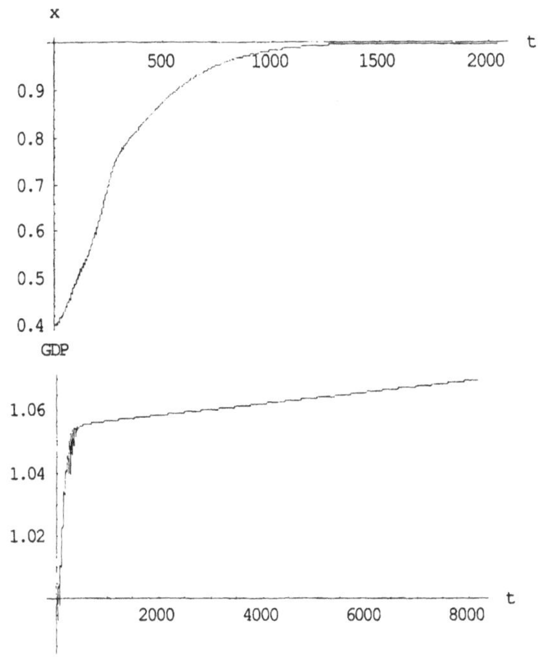


The second scenario represents the case where the knowledge stock chosen up to the previous period is not considered at all for both the real and financial sectors. This is the case in which the replacement of knowledge stock is promoted with full decpreciation of past investments because of no sunk cost for past investments, or their strategies or policy guidance. In this case (Fig. 3), fierce and highly irregular (chaotic) knowledge stock switches occur in both the real and financial sectors. As a result, knowledge stock that has been chosen at some stage is replaced by new knowledge stock even before it has had a chance to spread throughout the system as a whole, and the outcome is that the knowledge stock continues to depreciate over time without any benefit to the system as a whole. Therefore, at $t = 8000$, GDP growth is limited to around a 2.5% increase, which is lower than the first scenario. However, in this second scenario, it should be pointed out that locking into one type of knowledge stock can be prevented and conversions to new ones, that is technological changes, can occur.

**Fig. 3.** Knowledge stock selection ratio for the real sector and changes in GDP when previous-term earnings are depreciated in full

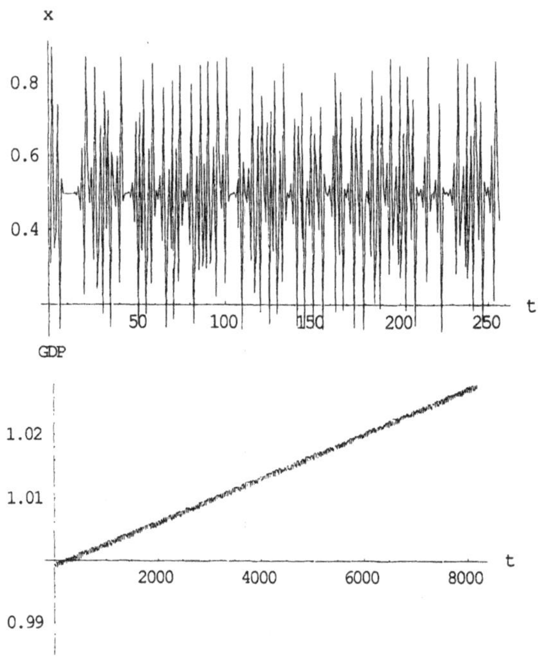


The third scenario involves calcuating the GDP that can be expected in both of the above two extreme scenarios, and controlling sustainable growth with knowledge stock shifts while changing the depreciation rate through learning. There are three characteristics in this third scenario.

In the first place, in this scenario (Fig. 4), by combining the economic welfare benefits of both scenario 1 and scenario 2 through managing the depreciation rate, there exists a path through which locking into one type of knowledge stock can be prevented and the results of technological innovation are spread throughout the national economy. At the same time, steady economic growth is achieved over the long term as conversions to new knowledge stock are induced. This path involves two phases: the process where technological change from existing knowledge stock to a new one is encouraged, and the other process where this new knowledge stock is fully disseminated through the whole economy while GDP rises sharply. Consequently, GDP will grow 2.2 times at $t = 8000$, which is much higher growth than seen in the other two extreme scenarios.

**Fig. 4.** Knowledge stock selection ratio for the real sector and changes in GDP when managed by the depreciation rate

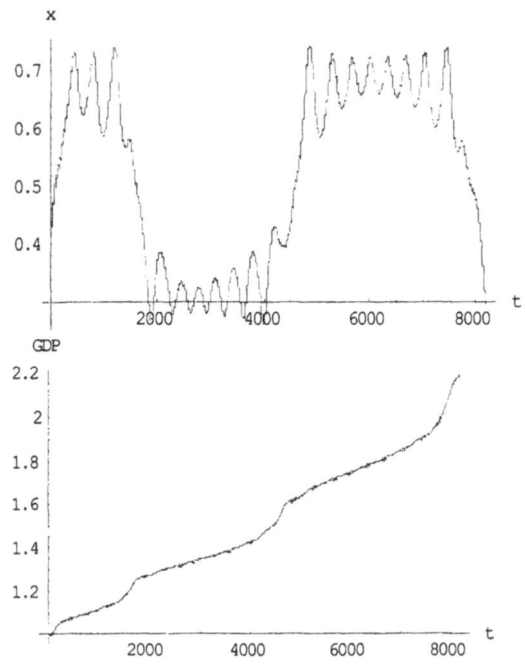


In the second place, the path for long-term growth through such technological changes is recognized as having the following characteristics. First, when one type of knowledge stock is standard throughout the entire economy, fluctuations are seen over time in the direction of knowledge stock selection, and over time, there is a decentralization of resource allocation away from the dominant knowledge stock and toward new knowledge stock. In other words, at a time when a particular knowledge stock forms the technological paradigm in the economic system, there is a push to "diversify and deviate" resource allocation to sources of technology other than the dominant one, so as to encourage a switch to new next-generation technology.

On the other hand, at the time when there is technological change from the existing knowledge stock to other new knowledge stock, over time the selection ratio moves more or less uniformly, and the fluctuations and instabilities that are seen during times when a particular technological paradigm is dominant do not occur. In other words, during the process of switching from technology that constituted the existing paradigm to new technology, resource allocation is more concentrated so as to promote the switch to next generation technology (that already has clear potential). That is, technology tends to become "standardized."

In the process where there is a move away from the dominant era of one particular knowledge stock in an attempt to convert to a new knowledge stock (for example, at around $t = 4000$), there exists a complicated choice as to whether to concentrate resources into existing knowledge stock that continues to spread throughout society but that has a relatively low marginal productivity, or whether to diversify into new knowledge stock that has an extremely low diffusion rate but remarkably high marginal productivity, and therefore there is no straight out switch. In other words, the switch to new technology seems to overlap with existing technology and is achieved after some going back and forth between the old and new technologies. In research and development competition amongst companies as well, technological development that will change the competitive framework, to a certain extent, is based on existing technology. This is due to the competitive pressure generated by the new technology and the sunk costs in the existing technology. As a result, it is not always the case that all companies make the one-way switch to new technology, but rather, companies generally go through the process of oscillating between the old and new technologies due to competition between the two, before completely switching to the new technology.

Finally, in the above process of switching and dissemination of knowledge stocks, the depreciation rate follows the following path shown in Fig. 5. More specifically, the GDP growth rate is relatively high at the time of knowledge stock changeover, but the depreciation rate at this time is relatively low (for example, at around $t = 1500$). Moreover, when a particular stock is becoming widespread, the GDP growth rate falls, and the depreciation rate at this time is relatively high (for example, at around $t = 2500$).

**Fig. 5.** Changes in the depreciation rate. The plot on the lower shows the depreciation rate plotted against GDP with a standardized scale of [0,1]

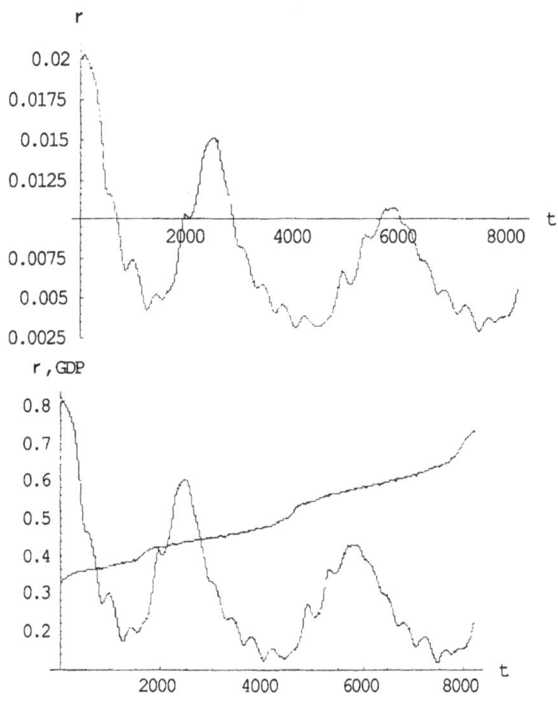


## 2.3.3 Policy Implications

Given the above findings, we now summarize the systemic (institutional) policy issues for encouraging sustainable growth with technological innovation from the perspective of controlling the depreciation rate. First, there exists a path in which depreciation rate management can prevent locking into a given knowledge stock and can stimulate the spread of the fruits of technological innovation throughout the whole national economy. This simultaneously sparks diversification and switches to new knowledge stock, while sustainable economic growth is still achieved over time. As such, there are some conditions for this sustainable economic growth with technological change (switches to new knowledge stock). That is, the distribution of earnings generated by research and development activities must show a heavy bias toward a reasonable concentration of high productivity amongst a small number of research and development agents (Sakaki 2004a; Scherer 1999).

In this case, there are two GDP growth triggers. The first trigger acts to increase "static efficiency," which boosts GDP by spreading a particular technology throughout the system. However, even if a GDP boost is achieved in the short term, this action on its own can cause a locking into this technology (knowledge stock field), causing GDP to stagnate from this point onward. To prevent such lock-ins to any particular technology, a second trigger is required that works to increase "dynamic efficiency," which sparks the switch to a new, more productive technology once the existing technology has become widespread. In this chapter, the "depreciation rate" functions as these triggers to manage policies indirectly through learning. That is, as a particular technology starts to spread throughout the entire economic system, keeping the depreciation rate relatively low will work to promote the accumulation of that knowledge stock. On the other hand, as a particular technology spreads throughout the entire economic system and productivity starts to fall, keeping the depreciation rate relatively high will act to promote depreciation of that knowledge stock and prevent lock-ins.

Moreover, when the knowledge stock depreciation rate is controlled according to the above conditions to form a sustainable GDP growth path, the depreciation rate time series showing the proportion of resource allocation to encourage dynamic efficiency for the whole system, indicates that this allocation remains at very low levels over the long term. That is to say, even the slightest degree to heterogeneity (agents that are inefficient and unstable in terms of static efficiency, but that have a strong orientation toward new technology in terms of dynamic efficiency) built into the system can enhance the performance of the whole system over time. In other words, by incorporating a primary policy objective which makes increased incentives for the real and financial sectors to maximize the use of existing knowledge stock accumulated to date, it can boost economic growth in the short term. Meanwhile, by incorporating a subordinate policy to encourage and support entrepreneurs focusing on the marginal performance of new knowledge stock, sustainable growth can be maintained by preventing lock-ins to particular technologies over the long term.

## 2.4 The Japanese Economy and Sustainable Growth

When we look at the relationship between technological change management and production benchmarks over time, provided that the results generated by the simulation are correct, we see that when marginal productivity of knowledge stock is relatively high, the depreciation rate falls. Conversely, when the marginal productivity is relatively low, we see an increase in the depreciation rate. In this way, sustainable growth can be achieved through technological change. This section focuses on Japan's asset securities reports, as well as private sector capital stock statistics to (1) consider the two elements of technological development paths: standardization and diversification, and (2) make an empirical examination of the success of technology change management in Japan's postwar economy.

### 2.4.1 Diversification and Standardization in Technological Development

Generally speaking, in a world where there are various possibilities for technological development, there are times when the focus is on technology diversification and times when standardization is promoted. According to Sudoh (1995), "Technology has the potential to grow in multiple directions, but when there is increased investment in dominant technology and massive investment in capital equipment and know-how; managers tend to be reluctant to choose to invest in alternative technologies. As a result, there is a tendency towards standardization in technological development over the long term."

Our simulations on this occasion produced results that correspond to his observations (Sakaki 2004a, 2004b). That is, throughout the period of changing from one knowledge stock to another (even after switching back and forth between the old and new), a more or less unilateral move toward one knowledge stock is observed. Conversely, when one knowledge stock has a dominant position, the share of that stock is seen to fluctuate over time. Therefore, in such a path of technological development, the two phenomena of diversification and standardization are likely to alternate over time, with diversification apparent in times of stagnant marginal productivity (or growth rates) and standardization occurring in times of high marginal productivity.

In this section we examine the above hypothesis based on data from asset securities reports for companies in the manufacturing industry listed in Section 1 and Section 2 of the Tokyo, Osaka, and Nagoya stock exchanges in Japan, obtained from the Development Bank of Japan's Corporate Finance Data Bank. First, we define "the rate of technological development orientation" as follows.

Rate of technological development orientation

= R&D expenditure/ (R&D expenditure + plant and equipment investment) ×100

Here we use "experimental research expenses" as R&D expenditure. It can also be considered as a benchmark for the focus on new technological development given the time lag from when the investment is made to when concrete results are seen in term of returns. On the other hand, plant and equipment investment can be seen as a benchmark for the focus on existing technological development, given that the existing level of technology produces actual machinery and equipment through this expenditure.

The rate of technological development orientation calculated for individual companies is rated from 0% to 100%, where the larger the figure, the greater the focus of that company on new technological development, and the smaller the figure, the greater their focus on the development of existing technology. This rate is obtained from all companies (those where the figure is available) each business year, and the standard deviation is calculated. This standard deviation shows the spread across all companies in terms of their orientation to technological development (either focusing on new or existing). As a result, when there is a large standard deviation in the rate of technological development orientation in any given year, this shows the disparity amongst companies in terms of their approach to development, that is, in the proportion of resources they allocate for "new technological development" and "existing technological development," respectively. In other words, times when the standard deviation is large can be taken to be times of technology "diversification," because there is a large difference in the proportion of investment allocated between new and old technological development. Conversely, when there is a small standard deviation in the rate of technological development orientation in a given year, this indicates that companies are taking more or less the same approach in what proportion of resource allocation they decide between new and existing technological development. These can be considered the times of technology "standardization," because there is little disparity between companies in terms of their investment in technological development.

We examined the relationship between the standard deviation for the rate of technological development orientation over the 34 years from 1969 to 2002 and the marginal productivity of knowledge stock (refer to Sect. 2.4.2 for the estimation techniques used) where there is a time lag of one period in the standard deviation for the rate of technological development orientation.

To investigate the relationship between standard deviation for the rate of technological development orientation and marginal productivity, we use a regression model where the former element is represented by $y$ and the latter by $x$. In this time series data, there is a positive serial correlation. Therefore, we calculate the coefficient of $x$ using the generalized least-squares method that accounts for first-order autocorrelation of the error term. The result is as follows (the $t$ value in brackets).

$$y = 22.83 - 0.419x$$
$$(10.72)\ (-4.43)$$
$$n = 33, \mathrm{SE} = 1.13, R^2 = 0.79$$

($R^2 = 1$ − residual variation / variation of $y$)

The results showed a negative regression coefficient (−0.419) and are significant at 0.1% ($t$ value=−4.43). Given that the regression coefficient is a significant negative correlation, after a one-period lag, it signifies a relationship in which when the marginal productivity of knowledge stock falls, technology is diversified, and when marginal productivity rises, technology is standardized (see Fig. 6).

**Fig. 6.** Standard deviation in the rate of technological development orientation and changes in marginal productivity for listed manufacturing companies in Japan.

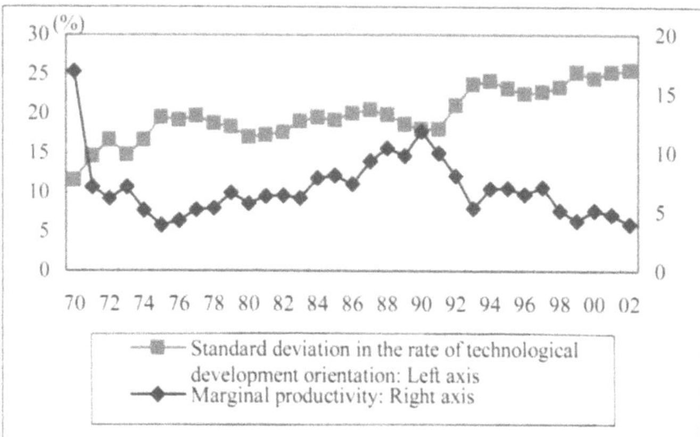


Source: prepared using the Development Bank of Japan's Corporate Finance Data Bank

### 2.4.2 Marginal Productivity of Knowledge Stock and Depreciation Rate of Capital Stock

In this section we estimate the marginal productivity of knowledge stock based on Hall's framework (Hall 1998). Knowledge stock $K_{it}$ is measured by the cumulative total of "experimental research expenses" for each period and each start-of-term balance that has been discounted at a set rate. This discount rate (rate of obsolescence) follows Hall's 15%. Next, capital stock $A_{it}$ provides the productive enterprise value of stock other than knowledge stock, and is defined as the sum of the "machinery and equipment" ending balance for the current period from property, plant, and equipment (i.e., tangible fixed assets), and the ending balance for the current period for "intangible fixed assets," which include goodwill, patent rights, software, etc. The knowledge stock and ordinary capital stock defined above make up the Cobb-Douglas production function in which firm market value $V_{it}$ is found for the factors of production. A simple linear regression model can be used to estimate the elasticity of the firm market value to knowledge stock balance $\alpha_t$ and then we can calculate the marginal productivity of knowledge stock.

$$\log\frac{V_{it}}{A_{it}} = \log q_t + \alpha_t \log\frac{K_{it}}{A_{it}}$$

We have used the same data as in the previous section: asset securities reports of listed manufacturing companies from the Development Bank of Japan's Corporate Finance Data Bank. Marginal productivity is calculated each business year using a cross section of data for companies that fit the description in this section.[2] Then the capital retirement rate is calculated based on the tangible fixed asset balance (based on acquisition cost) and retirement values for companies selected according to a common criteria. Then the capital retirement rate is substituted as the depreciation rate for existing knowledge stock[3] and the changes are plotted against the marginal productivity calculated earlier for comparison.

Accounting data for marginal productivity is available from 1969 onward, and for capital retirement rates, from 1977 onward, and the results are shown in Fig. 7. After falling rapidly throughout the end of the high-growth period and the 1973 oil shock, marginal productivity withstood the second oil shock and the high-yen recession. It then increased dramatically in the late 1980s when the Japanese economic system, characterized by the just-in-time production system, swept the

[2] All of the regression coefficient $\alpha_t$ throughout these periods are significant at less than 0.1%.

[3] Tangible fixed assets can be seen as a benchmark for the focus on existing technological development, given that the existing level of technology produces actual machinery and equipment through this investment.

world. However, marginal productivity then fell rapidly in the early1990s at the latter part of the bubble economy, and then fell again in the late 1990s. Whereas Japanese listed manufacturing companies kept the capital retirement rate of tangible fixed assets like production capital stock at a relatively low level during the 1980s when marginal productivity was rising and made optimal use of this production capital stock to maximize production capacity, in the late 1990s when marginal productivity fell again, they increased the capital retirement rate and switched to new capital stock.

Therefore, it is noteworthy that Japan's listed manufacturing companies satisfied the conditions for managing sustainable growth as suggested by the simulation results, from the late 1970s to especially the economic slumps in the late 1990s. In the United States at this time, there was already growing interest amongst some economists in theories for the "new economy" — the IT economy, while in Japan, as for these companies mentioned above, they succeeded in making the switch to IT economy-based knowledge stocks.

**Fig. 7.** Changes in marginal productivity and capital retirement rates in Japan's listed manufacturing companies

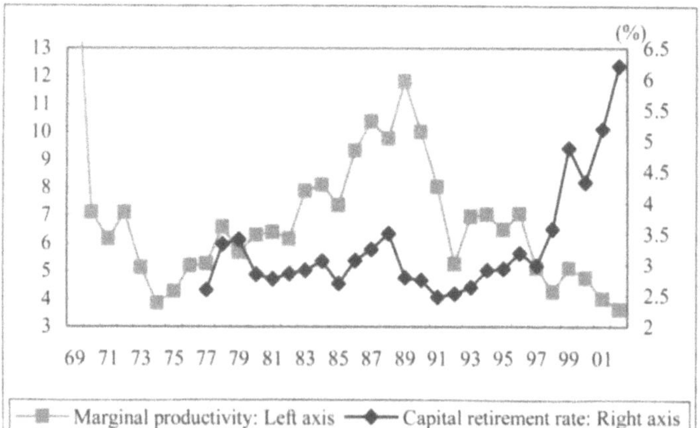


Source: prepared using the Development Bank of Japan's Corporate Finance Data Bank

### 2.4.3 IT Paradigm Change and Systemic Inconsistency in the Japanese Economy

Now that we have entered the new millennium, why is it that the Japanese economy is still plagued by recession and why are there still no signs of productivity increases that would indicate a technological change that embraces the so-called IT economy? Figure 8 shows the changes in the 1990 and 1995 constant price capital retirement rates based on acquisition prices across all manufacturing companies using private sector capital stock statistics in addition to the retirement rates for listed manufacturing companies dealt with earlier. A comparison of the capital retirement rates in these listed manufacturing companies and macrolevel manufacturers throughout the entire Japanese economy highlights the following points.

**Fig. 8.** Changes in the capital retirement rates of listed and nationwide manufacturing companies

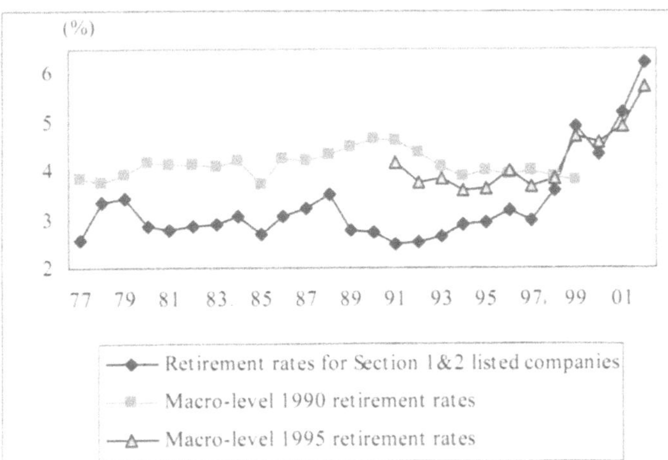


Source: prepared using the Development Bank of Japan's Corporate Finance Data Bank and Economic Social Research Institute, Cabinet Office's private sector capital stock statistics

First, the macrolevel capital retirement rates in 1990 constant price; which are available over a long time series, and those of listed companies show quite different changes over time. As was pointed out earlier, in the 1980s when marginal productivity was rapidly increasing, listed manufacturing companies kept their capital retirement rates for tangible fixed assets at relatively low levels and made optimal use of their production capital stock to maximize their production capacity. In contrast, when marginal productivity fell in the latter part of the 1990s, they increased capital retirement rates and converted to new capital stocks. However, the 1990 constant price macrolevel capital retirement rates seem to have moved in the opposite direction to those of listed companies in that they were gradually rising from the late 1970s to late 1980s and then went into a declining trend in the 1990s.

Second, the capital retirement rate time series for macrolevel manufacturers, in both the 1990 and 1995 constant price, show less change over time than the capital retirement rate for listed manufacturing companies. This means that manufacturing companies as a whole national economy may not have taken a flexible approach to the promotion of capital stock diffusion and conversion. Admittedly the capital retirement rates for macrolevel manufacturers in 1995 constant price seem to be increasing from 1999 to 2002 keeping step with those for listed manufacturing companies. But whereas the capital retirement rates of listed companies showed roughly a twofold increase (2.4 percentage points) from the early to late 1990s, manufacturing companies on a national scale saw only an increase of 1.1 times (just over 1 percentage point) by 1995 constant price. When compared with the 1990 constant price data, this shows that there was a declining trend in their capital retirement rates throughout the 1990s.[4]

It has also been pointed out that one contributing factor to the information technology productivity paradox is inefficiency that causes only partial implementation of information technology. If a system within a company is to

[4] Marginal productivity time series for macrolevel manufacturers cannot be estimated because of lacking cross section data of all manufacturing companies. However, we can calculate GDP to plant and equipment investment ratio instead of marginal productivity, which represents revenue based on value added gained per unit of real productive investment. There seems to be an opposite correlation between the GDP to plant and equipment investment ratio and the capital retirement rate time series for macrolevel manufacturers from the early 1950s to the early 1990s. See Sakaki (2004a).

generate sufficient cost reductions, the information technology needs to be implemented thoroughly throughout all processes and organizational departments in that company's system. If this information technology is only partially implemented, then this will only serve to prevent those sections not updated from achieving cost reductions, and if anything, trying to maintain connectivity with these processes and sections not updated will reduce the company's overall IT performance (Lillrank and Sudoh 1998).

There are 2346 companies listed on Sections 1 and 2 of the Tokyo, Osaka, and Nagoya stock exchanges in Japan registered in the Development Bank of Japan's finance data used for this study, and 1327 of these are manufacturing companies (based on end of FY2000 data). Meanwhile, according to the FY 2001 Establishment and Enterprise Census conducted by the Ministry of Public Management, Home Affairs, Posts and Telecommunications, as of 1 October 2001, there were 1 617 535 companies in Japan, and 297 510 of these were manufacturing companies. No more than 0.15% of all companies are listed companies, and only 0.45% of manufacturing companies are listed companies. In over 99% of cases, transactions between these unlisted companies that have been slow to adopt IT and networking technology and the listed companies with progressive IT systems are fraught with inefficiencies, including problems like not being able to conduct electronic transactions. That is, business dealings with unlisted companies not only impede the diffusion of e-commerce, but also require that the conventional business mediums like facilities such as the telephone and fax, systems and organizations for those personnel handling the transactions, and conventional business conventions be maintained. This could quite possibly be increasing the burden of inefficient investment and operations costs not only for listed companies, but for society as a whole.

In actual fact, according to the section of the census that looks at the state of e-commerce, 169 000 companies have introduced e-commerce, which equates to a mere 10.5% of all companies. Also, if we look at e-commerce adoption rates according to levels of capitalization, we find that e-commerce is used in less than 10% of companies capitalized at 10 million yen or less, which account for no less than 50% of all companies. These small-capitalization companies account for half of all companies. By contrast, 44.4% of companies capitalized at over 5 billion yen are using e-commerce, which account for no more than 0.1% of all companies. These figures show that the use of e-commerce is largely dependent on the level of capitalization.

## 2.5 Conclusions

Today, Japan is grappling with making the switch to knowledge stocks that are in line with the increasing use of IT, and there is the possibility that systemic inconsistencies in the IT push in terms of facilities, the legal system, and organizational structures will cause a protracted recession (Freeman 1987, 1992). Given this situation, even if some listed companies with increasing capital retirement rates promote steps to convert knowledge stocks, we may still not be able to enjoy the increased performance that the switch to IT could generate, because unlisted companies, which form the broad base of the Japanese economy, are slow to adopt IT measures. This alone is enough, but coupled with the inconsistency between the systems that govern the knowledge stock for listed and unlisted companies, has the potential to further harm productivity, meaning that these mismatched systems are highly likely to generate even greater inefficiencies ("systemic-inconsistent inefficiency").

## Acknowledgments

The author expresses his sincere gratitude to Dr. Osamu Sudoh (Professor, The University of Tokyo), Dr. Hiroshi Deguchi (Professor, Tokyo Institute of Technology), and Dr. Gyoichi Iwata (Former Professor, Keio University) for providing valuable comments. This research was supported by a Grant-in-Aid for Specified-Area Research, "Informatics Studies for the Foundation of IT

Evolution" which is funded by The Ministry of Education, Culture, Sports, Science and Technology.

## References

Deguchi H (2000) Economics as complex systems (in Japanese). Nikkagiren, Tokyo

Dosi G (1990) Finance, innovation, and industrial change. Journal of Economic Behavior and Organization, 13:2002, pp 299–319

Hall BH (1998) Innovation and market value. NIESR Conference on Productivity and Competitiveness, London, February 5–6, 1998

Freeman C (1987) Technology, policy and economic performance (in Japanese). Koyo Shobo, Tokyo

Freeman C (1992) The economics of hope. Pinter, London

Lillrank P, Sudoh O (1998) Software society (in Japanese). Fujitsu, Tokyo

Sakaki S (2004a) Research into sustainable economic growth through knowledge stock management: an explanation of the mechanisms for technological change using a simulation model and an empirical study of the Japanese economy (in Japanese). Doctoral Thesis, The University of Tokyo, Graduate School of Humanities and Sociology

Sakaki S (2004b) Management model for technological change and sustainable growth. Evolutionary and Institutional Economics Review 1.1

Scherer FM (1999) New perspectives on economic growth and technological innovation. Brookings Institution Press, Washington DC

Sudoh O (1995) Interactive network society (in Japanese). Yuuhikaku , Tokyo

Yoshikawa H (2000) Modern macroeconomics (in Japanese). Sobunsha , Tokyo

# 3 Capital–Skill Complementarity in the United States and Japan

Reiko Gotoh

Faculty of Humanities, Ibaraki University, 2-1-1 Bunkyo, Mito 310-8512, Japan, rgotoh@mx.ibaraki.ac.jp

**Summary**

Productivity-paradox, widening wage differentials, and increasing insecurity among industrialized nations throughout the 1980s and 1990s call for a closer look at the impact of information and communication technology on employment. This chapter investigates the impact of the United States and Japan in this area. There are two competing hypotheses regarding skill-biased technological change and capital–skill complementarity. Based on empirical studies and analysis using firm-level data, this chapter discusses a linkage between the two hypotheses and shows that the absolute demand shift is more important than the relative demand shift. The theoretical and political implications of these findings are discussed.

**Key Words** Capital–skill complementarity, Skill-biased technical change, Job security, Wage differential, Diversification of employment patterns

## 3.1 Information and Communication Technology, Organizational Capital, and Skill

### 3.1.1 Intangible Assets

The so-called productivity paradox frequently emerges during discussions of the relationship between investment in information and communication technologies (ICT) and productivity. The productivity paradox refers to the phenomenon in which there is no positive correlation, at least in macro level statistical terms, between an investment in ICT and productivity, in contrast to physical investment in machinery or facilities. Yet numerous OECD countries including Japan have already confirmed that ICT investments contribute to productivity.[1] The focus of discussions has shifted to "Under what situation does ICT enhance productivity, and by how much?" or, "What is the nature of the linkage that connects ICT and productivity?."

A focus on the link that connects ICT investment and growth in productivity has stimulated interest in a fact that has been repeatedly raised in former studies on incorporating factory automation (FA), namely, that there is a correlation between automation technologies, organization structures, and employment relationships.[2] Recently, the correlation has been often referred to as the "complementary relationship between ICT capital stocks and intangible assets."[3] This means that ICT investments bring a significant rise in productivity only when

[1] See, e.g., Jorgenson and Motohashi (2004). They suggest that when the national accounts for Japan and the United States are harmonized, the contribution of IT to economic growth turns out to be nearly identical. The differences of economic growth rate of the two countries come from labor inputs and non-IT capital input contribution to economic growth.

[2] At early stages of IT development, NC working machines and flexible manufacturing systems (FMS) were often adopted at processing and assembly-type manufacturing sites. For this reason, the adoption of IT during this phase is often referred to as factory automation (FA).

[3] "Complementary" means that the value of one thing is further enhanced by the value of the other. There is a growing number of literature reports on intangible assets. See, e.g., Hand and Lev (2003).

accompanied by complementary intangible assets. According to Brynjolfsson et al. (2002), organizations that have attained high market value and productivity have invested in computer hardware and intangible assets at a ratio of 1:9. In other words, investments in ICT capital stocks account for only a small portion of the total investment.

What, then, are these intangible assets that are in a complementary relationship with ICT? According to the analytical framework presented by Kaplan and Norton (2004) and Bresnahan et al. (2002), which are based on firm-level quantitative and qualitative data, the intangible assets that support the knowledge society can be broadly categorized into: (1) information capital, (2) organizational capital, and (3) human capital. Information capital refers to the ICT and digitized work processes that enhance productivity of workers; organizational capital refers to the organizational capability to enhance the productivity of workers, and includes organizational structure, task allocation, decision-making distributions, relations with suppliers and major customers, and the culture of the company; human capital refers to the labor force characteristics of the knowledge, skills, innovativeness, and adaptabilities to change.

In generally, firm-level data allow us to go behind published statistics at the macro level or industry level. Firm-level data studies suggest that businesses need to make parallel investments in organizational capital and human capital with ICT investment to yield productivity gains. In other words, ICT use, revision of workplace structures and practices, and highly skilled labor and investment in training are complementarities.

### 3.1.2 ICT and Organizational Capital

If intangible assets are the critical assets for organizations today, the value of organization capital will increase to include the knowledge of all employees being put to work. Therefore, organizational capital—which is in a complementary relationship with information capital—is believed to substantially overlap with the so-called Japanese style management system, which features a flatter organization, decentralized decision making, self-managed teams, job rotation, and so on. Such work systems are called "high performance work systems" or "knowledge-based work systems" (Osterman 1999; Osterman et al. 2001).

Looking back in history, it was during the 1970s that ICT began to be fully incorporated into the work places of developed countries. In a typical internal labor market prior to IT development, the distribution of the labor force and incomes were determined beforehand based on detailed hierarchical job rankings (Doeringer and Piore 1971). Worker productivity was determined primarily by job allocation in the internal labor market, rather than by the capabilities and efforts of individuals. Consequently, the costs of metering team productivity could be reduced by having bureaucratic managers acting as monitors (Alchian and Demsetz 1972).

Since the 1970s, however, market competition and uncertainty have expanded significantly. Organizations were left with no choice but to substantially reform the organizational structures and employment relationships. The need for reform was felt most strongly in countries like the United States, France, and the United Kingdom, where labor-management relations have traditionally been confrontational and decision-making processes depend on vertically centralized structures. While corporations in these countries attempted to cut labor costs by precluding labor unions and by giving priority to downsizing and reducing wages, corporations in Japan, Germany, and Italy, where employment relations have tended to be relatively cooperative, have tried to improve quality and productivity by setting mutual gains as a common goal for labor and management. This has been performed by exchanging high-level employee commitment and flexibility of the internal labor market for long-term training and employment security (Piore and Sable 1984).

After the downsizing and reengineering of the 1980s and 1990s, the well-defined and vertical organizational structure of corporate enterprises is disappearing, even in the US economy. In the United States, the adoption of knowledge-based work systems was not common in the early 1990s (Gordon 1996). It was after the mid-1990s that the systems began to reach the expanded workplace of managerial, clerical, expert, and other white-collar workers. It was also after the mid-1990s that ICT-linked growth in productivity started to be seen in the overall economy.

According to O'Reilly and Pfeffer (2000), even in the world of high technology and software development, there is a case to be made for treating people differently and better than they expect to betreated, and achieving extraordinary results with ordinary people – to lock the value hidden in all organizations and in all people. New United Motor Manufacturing Inc. (NUMMI, an automobile manufacturing plant run by General Motors and Toyota that started operation in 1983 as a joint venture), which the authors cited as an example of an employee-centered organization that has attained favorable outcomes, is a good example. It shows that investing in better manufacturing processes and in reforming human resource management methods, rather than in automation technologies and special workforce, can be a more effective way of bolstering productivity (Scott Morton 1991).

Labor economy experts may believe that NUMMI is an obsolete example. However, the organizational system, in which workers are given autonomy by sharing a broader range of information and information processing efficiency is improved in a decentralized and horizontal way, is complementary with current ICT development. Two empirical studies using detailed firm-level data, Bresnahan et al. (2002) and Economic Planning Agency (2000), enable us to empirically assess the correlation between ICT and workplace organization in the United States and Japan. They found that the productivity of firms with high ICT investment and high organizational investment to help decentralize workplace organization was greater than that of firms in which both or either of these investment types was low. The decentralized workplace organization variable is aggregated as a set of various measures on team-based work organization and individual decision authority, and the data set contains not only productive plants but also nonproductive organizations. These studies suggest that knowledge-based work systems are applicable to both blue-collar workers and white-collar workers.

### 3.1.3 ICT and Human Capital

Regarding complementarity between ICT and human capital, many empirical studies show that ICT is "skill-biased technical change" (SBTC).[4] SBTC means that recent technological change in production techniques, ICT capital deepening, and organizational changes are the driving forces of growth in the relative labor demand shift from unskilled workers to skilled workers. There are two types of studies that suggest that the hypotheses of SBTC explain the wage gap in OECD countries. One sets a focus on wage differentials by observed skill attributes. For example, Autor et al. (1998) suggest that the recent widening of the US educational wage differentials has been led by an effect of SBTC as measured by computerization. The other studies set a focus on a sharp growth in within-group (or residual) wage inequality (e.g., Bartel and Sicherman 1999).

The purpose of the work by Bresnahan et al. (2002) was to sharpen hypotheses about SBTC and correlation between ICT and skill. They argued that three complementary innovations, (1) the increased use of ICT, (2) changes in organization practices, and (3) changes in products and services, constitute the SBTC that calls for a higher-skilled labor mix. Factors that combine the three innovations are "limited substitution" and "information overload." Technologies are unable to substitute for all businesses. The more that typical work tasks and routines are substituted by ICT, the more that added value is generated from human skill or soft skill. In addition to that, digitized work processes need many information-processing tasks. Therefore, firms must enhance lateral communication efficiency to decentralize workplace organization, hire high-skilled workers, and give them autonomy.

Bresnahan et al. (2002) and Economic Planning Agency (2000) also investigated the correlation between ICT and human capital in the United States and Japan. Measures of human capital are a set of variables on human capital levels (e.g., education mix and occupation mix) and human capital investment

[4] See, e.g., Berman et al. (1998) and Autor et al. (1998) for the United States, Ohara and Ohtake (2001) and Harashima and Tejima (2002) for Japan, and Mashin and Reenen (1998) for OECD.

(e.g., training, cross-training workers). They also confirmed the complementarities among high IT investment, decentralized workplace organization, and high human capital levels/ human capital investment.

However, according to Harashima and Tejima (2002), no evidence has been found in Japan to support the hypothesis of SBTC from surveys on wage gaps based on observed skill attributes. Wage distribution gaps among job ranks, professions, education, and occupation remained almost constant throughout the period 1985 to 2000, and even showed a tendency to level off.[5] In contrast, a growth in within-group wage differentials is supported in Japan as in the United States (Ohara and Ohtake 2001). Why have wage gaps based on observed skill attributes not been widening in Japan? To explain the reason, it is useful to sharpen the hypotheses about SBTC.

## 3.2 Relative Demand Shift from Deskilling to Upskilling?

### 3.2.1 Capital–Skill Complementarity Versus SBTC

One of the important technical features of ICT is that high marginal productivity is exhibited when large volumes of routine information that could be converted into formative intelligence are batch-processed at high data rates. The feature makes it theoretically inevitable that the demands on skills for routine processing decreases, and demands on skills for processing nonroutine information increases. Given that today more and more people are being required to have skills in nonroutine processing of information, recent technical change is definitely a form of SBTC. It is difficult to deny the hypothesis of capital–skill complementarity.

However, the complementarities between ICT and intangible assets that are built up by knowledge-based work systems and higher human capital investment

[5] They use *The Basic Statistic Survey on Wage Structure* for the period 1985 to 2000, making adjustments for age differences, while not all attributes have been adjusted rigidly as in Autor et al. (1998).

are not very consistent with the widening wage differential based on visible skill attributes, while it may explain a growth in within-group wage differentials.

The existence of complementarities between ICT and intangible assets signifies that through ICT deepening and decentralized workplace organization, the value of noneconomic assets that is difficult to reduce to an economic value and traded in the markets has become relatively higher. In other words, the asset value of "firm-specific skills" in human capital theory or "relation-specific knowledge" that differ from those possessed by other organizations are more important than ever before.

If intangible assets are critical to gain high profit from ICT investment, firms are willing to jointly invest in relation-specific knowledge, and wage increases may increase productivity. This hypothesis is called the efficiency wage theory. If this hypothesis of labor and wage–productivity dependency is appropriate, there is positive correlation between ICT investment and not attributes of individuals' skill but rather firm-specific attributes. The speed of employment adjustment will slow and employees will obtain job security if firms pay more expenses for specific training of employees.

O'Reilly and Pfeffer (2000) noted the following: "The payoff function is not linear but exponential. Firms that invest in people but don't have systems that encourage long-term employment and career development are likely to reap reward from their investment." At ICT companies in the Silicon Valley, job tenure is generally short and the propensity of risk to workers is great, so strong performance-based incentives and stock options are provided to prevent the core human resources from leaving the company. Still, employee-centered, high-performing companies have achieved extraordinary results by bolstering employee satisfaction and loyalty through incentives encompassing growth, teamwork, challenges, and remuneration for achievements, rather than with high wages. Companies that have remained highly competitive in the high-tech sector, where competition is fierce to acquire skilled workers, have chosen to raise productivity not through lay-offs and wage reductions but by inducing the "hidden value" out of their employees, and by supporting in-house training rather than spot-trading skills for high prices as commodities.

Moreover, the ICT-intensive firms that have enjoyed high productivity usually invest a great deal of time and money in training staff they have hired, even though most firms adopt pay-for-performance. These organizations emphasize screening procedures at the time of recruitment about not only the educational background and computer skills but also the cooperativeness and communication skills. Due to increased earnings and employment security, those firms gain good effects on turnover, job satisfaction, and commitment (Brynjolfson et al. 2002).

### 3.2.2 Flexibility Versus SBTC

As previously stated, many empirical studies in the United States show high premiums for individuals with high visible skills in coping with the changed environment. That can be interpreted as growth in relative demand for "general skills" or "visible knowledge". In other words, there are the complementarities between ICT and tangible assets.

Even though it is difficult to draw a clear line, from the perspective of transaction costs, the decentralized workplace organization makes the internal cost of monitoring and measuring, on the other hand, digitalization and networking cheapen external transaction cost of screening. And from the perspective of human capital theory, the boundaries among firms and the speed of knowledge and obsolete skills do not directly affect cost efficiency between internal resource allocation and external resource allocation. If technological change is not radical but continuous, firms should be motivated to ensure long-term job security and to jointly invest in the long-term training for employees. This is because it is more cost-efficient to pay in retraining expenses in the internal labor market than layoffs and buying in the external labor market, considering sunk cost. In the processes of "creative destruction" or "disruptive innovation" (Christensen 1997), dynamic resource allocation is more cost effective than internal reallocation because general adaptability to change is needed. Some have hypothesized that a market-type labor market like that in the United States is more effective in the information age, because its regulatory and institutional environment permits firms to reallocate human resources over the boundaries among firms.[6]

In such cases, the conclusion reached was that the old practices in workplaces will be substantially revised to become more flexible (Cappelli 1999). If the allocation of labor and the determination of wages within the internal labor market are close to the conditions of external labor markets, workers are likely to select the working style to suit which their career while moving from job to job to seek better conditions. The change in human resource management by firms and the change of workers' behaviors emerge as a spiral. The more flexible that the labor market becomes, the more incentives are needed to attract higher-skilled workers, and the cost of turnover increases. As a result, many risks of the labor market are imposed on workers. Employment patterns become more diverse and wage gaps widen between highly-skilled workers and other workers. Such circumstances well explain reducing job securities, widening wage differentials by observed attributes of individual skill, and displacement of unskilled workers who are the main component of the unemployment pool.

### 3.2.3 Relative Impact and Absolute Impact

Once again, the hypothesis of capital–skill complementarity appears to be valid, but the complementarities between ICT and intangible assets basically conflict with the wage differentials by visible skill and decreasing job security.

In this regard, there seem to be two hypotheses: one is that widening wage differentials according to visible skill and decreasing job security is a temporary phenomenon in a transition period. At times of revolutionary structural change, the cross-functional expert qualifications surpassing the functional expert qualifications of the past become necessary. If the demand for skill stabilizes, long-term investment in training for employees may result in human capital

[6] See e.g., Haltiwanger et al. (2003) for manufacturing sectors in Germany and the United States, and Nicoletti and Scarpetta (2003) for OECD countries.

investment costs that are lower than those incurred with layoffs and downsizing. This scenario emerges when ICT capital, organizational capital for decentralizing the workplace, and human capital to improve skill levels of employees accumulate.

The other hypothesis is that the adoption of knowledge-based work systems is only a temporary strategy that some firms can adopt for some skilled workers in a short period. The hypotheses about SBTC have a lacuna in that only the *relative* impact of technological change is considered. If there is a long-term trend that the relative demand shifts from deskilling in the Fordism age to upskilling in the information age, labor-saving technological change may imply an *absolute* reduction of both the skilled and unskilled workforce (Vivarelli and Pianta 2000). Without upsizing, knowledge-based work systems only enforce most risks of the labor market upon workers.

We cannot predict which hypothesis will be realized in the future, but the first hypothesis seems to be true only when the absolute demand for higher-skilled workers has been increased and to be so in the future. In the new boom economy of the United States from 1995 to 2000, both labor productivity growth and employment growth were greater in the IT-intensive industries compared with the less IT-intensive industries. Average annual growth of full-time equivalent workers in the IT-intensive industries was 2.87%, compared with 2.59% in the less IT-intensive industries (Dumagan et al. 2003). Maintenance of current output with fewer workers or achieving greater output with the current workforce results in labor productivity growth. The fact that greater labor productivity growth ocurred with greater employment growth in the IT-intensive industries means that the sector created new added value and many highly paid jobs, although the sectors might contribute to job displacement of low-skilled jobs or repetitive tasks more than the less IT-intensive sectors. Competition for highly skilled workers became intense by the shift in absolute demand for a skilled workforce, but this has resulted in knowledge-based work systems to improve mutual gains for firms and workers. In addition, a spillover effect to the overall economy was created.

Wage differentials improved because of a shortage of both skilled and unskilled workforce (Mishel et al. 2004).

However, even during the economic boom, insecurity remained high. The firms that have implemented knowledge-based work systems request employee commitment, but they do not necessarily assure their employees of long-term job security (Cappelli 1999). Furthermore, the knowledge-based work system is not the only possible way to compete. As Osterman (1999) suggested, knowledge-based work systems can coexist with downsizing and slow wage growth.

Moreover, during the economic downturn of 2001, employment growth in IT-intensive industries declined dramatically. The growth of full-time equivalent workers in the sector was −2.73%, whereas it was 0.43% in the less IT-intensive industries. Average annual growth of labor productivity in the sector declined slightly. It was 3.79% in 1995–2000 and 3.10% in 2001. That is, the productivity growth of 2001 in the sector was mostly led by current output with fewer workers. Dumagan et al. (2003) suggested that some of this job shedding in the sector may be directly related to ICT, because most job losses occurred in management and office administrative support occupations, and many of these occupations are susceptible to labor-saving technologies. In addition, some studies indicate that structural job losses may be occurring in US employment, especially in IT-intensive industries or IT-producing occupations. Some suggest that noncyclical factors of outsourcing abroad called "offshoring" have influenced displacement of highly skilled, highly paid jobs as well as low-skilled, low-paid jobs (Cooke 2003; Groshen and Potter 2003).

There is a "high-road" strategy in which firms adopt ICT-intensive and knowledge-intensive workplace organization to improve productivity so that firms and employees share mutual gains in which employee commitment is exchanged for employment security. Lipietz (1992) criticized the high-road strategy because (1) it consists with "low road", otherwise they are compatible in an economy and (2) high-road scenarios assured by individual firms should not be stable. Although it does not happen that all workers work as free agents, increased mobility, in other words, increased insecurity and wage differentials, are facts of life today and are likely to be so in the future in the US economy (Osterman et al. 2001).

Furthermore, it is worth noting that there is a well-known critical difference between the employment systems of the United States and Japan. The Japanese employment system has depended on the high-road approach, with long-term job and wage security of core employees, including blue-collar workers, mainly assured by individual firms by the early 1990s. This also means that the cost of hiring core employees to firms and the cost of turnover of workers are more expensive in Japan than in the United States. Throughout the long-term recession, the Japanese employment system has changed. The next section examines changes in the labor market structure and how technological changes affect employment relations in Japan.

## 3.3 The Employment Impact of ICT in Japan

### 3.3.1 Diversification of Employment Patterns

Instead of increasing insecurity of the overall labor market as in the United States, the distribution gap between the internal labor market and the external labor market has expanded in Japan. The employment practices of long-term job security and seniority-based pay for full-time regular employees are more or less being maintained. For example, the main measures for employment adjustment of large companies are entry restrictions and restriction of overtime. As indicated by the *Industrial Work Situation Survey* in 2000, the top measure is "reduction/suspension of recruitment of new graduates" (52.6%), and the second is "overtime regulations" and "reduction/suspension of midcareer recruitment" (32.7%), although "voluntary retirements and layoffs" have increased. The job tenure of remaining regular employees is getting longer, and, even taking into account the fact that the turnover rate is lower in a recession than in a boom, it cannot be said that long-term employment security for regular employees has broken down entirely. In addition, the seniority-type wage profile is also being maintained. As mentioned above, skill premium is not widening among employees.

We need not have apprehensions if the labor market has not become increasingly bifurcated between insiders and outsiders. However, the demand shift from regular to irregular employment is underway. The number of regular employees decreased by about 4 million during 1997–2002, while, in contrast, the number of nonstandard workers increased: part-time employment increased by about 1.5 million, with temporary workers in part-time jobs increasing by about 0.9 million and part-time workers increasing by about 0.83 million in the same period (*Employment Status Survey*, Ministry of Internal Affairs and Communication). The number of freeters[7] increased rapidly from 1.82 million in 1990 to 4.17 millions in 2001 (*White Paper on the National Life Style 2003*, Cabinet Office of Japan).

The diversification of employment patterns is partly because of the supply-side factor that workers prefer a nonstandard work style because it provides them with flexibility. However, the main reason is that the high security and payment for insiders bring about a reduction in demand for regular employees and asymmetrical risk conversion to outsiders such as young people and secondary workers (Genda 2001). Although the percentage of nonstandard workers has been traditionally high in Japan among OECD countries, the diversification of employment patterns in the late 1990s has experienced a structural change. That is, the cyclical pattern in which the percentage of nonstandard workers decreases during economic recession has disappeared. The full-time employment rate in the United States overtook that of Japan in the period (see Table 1).

Nevertheless, the changing demand pattern is not reflected in wages. The wage gap in the hourly wage between regular employees and part timers has widened from about 30% in 1976 to about 44% in 2001 (including annual bonus) (*Basic Survey on Wage Structure*). According to Ohsawa (2002), who investigated a

[7] Freeter is a term used in Japan to describe a temporary part-time worker, generally for those younger than 35 years.

factor analysis of the wage gap for different forms of employment among female workers, wage differential due not to differences in skill attributes, such as education and number of years work experience, but to differences in form of employment are better able to explain the wage gap. In other words, if differences in skill attributes are proxy variables for the quality of human capital, there is a strong possibility that the wage gap between full-time and part-time workers arises from the different wage structure by the employment patterns, and not from marginal productivity.

**Table 1.** Full-time employment in the United States and Japan

| Country | 1995 | 1997 | 1999 | 2001 |
|---|---|---|---|---|
| United States | 76.6 | 77.1 | 78.1 | 78.4 |
| Japan | 80.8 | 78.6 | 76.9 | 74.3 |

Full-time employees: work more than 35 h/week, including executives.
Sources: *The Special Survey of Labour Force Survey* (Japan), *CPS Supplements* (USA), Takayama (2004) p.41

The issue here is how ICT affects employment arrangements in Japan. Most survey and empirical studies based on firm-level data during 2000–2003 suggested that increased use of ICT and related organizational changes have caused regular employees to be replaced with part-time workers and temporary workers because of labor-saving substitution effects, and increased mobility of the labor market. Dynamic resource allocation among firms and the need for general adaptability to change instead of firm-specific skills were cited as reasons for this change.[8]

However, few studies analyzed the correlations between the IT-intensive firms and employment arrangement patterns or prove its labor-saving effect based on firm-level data including both production firms and nonproduction firms. In the following section, we discuss the correlations between the IT-intensive firms and employment arrangement patterns, using individual slip data of the *Survey of Organizations Using IT* (hereinafter referred to as the JIL Survey) conducted in

[8] Gotoh (2004) provides a convenient overview of survey results.

December 2000 by the former Japan Institute of Labour (currently the Japan Institute for Labour Policy and Training).

### 3.3.2 Human Resource Arrangement in ICT-Intensive Firms

#### *3.3.2.1 Properties of Data Source*

The JIL Survey covered organizations throughout Japan with 300 or more employees. It had 1637 effective responses, giving an effective recovery ratio of 16.4%. Let us begin by confirming the basic attributes and the situation of ICT use at organizations surveyed in the JIL Survey. *According to the Statistical Survey on Business Establishments and Organizations in Fiscal 2001*, conducted by the Ministry of Internal Affairs and Communications, there were 11 985 companies in 2001 with 300 or more employees. Therefore, it is estimated that the JIL Survey received responses from about 14% of the major companies in Japan.

Manufacturing companies accounted for 35.2% of the companies covered by the JIL Survey. According to the *Statistical Survey on Business Establishments and Organizations in Fiscal 2001*, manufacturing firms accounted for 18.4% of all companies and 31.2% of companies with 300 or more employees. Thus, the number of manufacturing firms among the respondents was slightly higher than the national averages.

Next, we briefly confirm the features regarding the progress in ICT use. Of the companies that responded to the JIL Survey, 92.1% have adopted e-mail systems, 94.5% use the Internet, and 19.3% use e-commerce. In a survey conducted by the Rengo Institute for Advancement of Living Standards in May 2002, which was after the IT recession trough (January 2002),[9] the percentage of organizations using e-mail was 97.0%, slightly higher than in the JIL Survey. Looking at the *Survey on Trends in Usage of Communications* by the Information and

[9] Rengo Research Institute for Advancement of Living Standards: *Survey on Effects of IT on Jobs and Work Place Organizations* surveyed individual white-collar regular employees at private firms nationwide.

Communication Policy Bureau of Ministry of Internal Affairs and Communications, the rate of Internet use was 89.3% in 2000 for organizations with 100 or more employees, and 94.5% in 2001. The figure for 2001 coincides with that in the JIL Survey. Finally, the rate of adoption of e-commerce as revealed in the *Statistical Survey on Business Establishments and Organizations in Fiscal 2001* was 10.5% in 2001. Because the average capital of organizations surveyed in the JIL Survey was about 85.35 billion yen, we should look at the rate of e-commerce use among organizations with capital equal to or greater than 50 million yen and less than 100 million yen. The average rate for this particular category was 20.6%, which is similar to the rate of 19.3% observed in the JIL Survey. Thus, the respondents in the JIL Survey are deemed to accurately reflect the status of ICT use at major Japanese organizations as of 2001.

#### *3.3.2.2 Measures of ICT Use*

In the JIL Survey, three questions were asked regarding the progress in ICT use. They were: (1) PC dissemination, (2) ICT tool adoption and use, and (3) ICT adoption at each division. Point (2) was to identify the adoption of six tools, namely e-mail, groupware, the Internet, company websites, backbone system, and e-commerce. In this study, the companies that responded to all seven questions regarding progress in (1) and (2) were extracted as samples to establish indices that revealed the progress in ICT use, as at The Japan Institute of Labour (Anon. 2001). Readers are referred to the document for details, but a brief explanation is provided here.

First, the following measures were taken to replace qualitative variables with quantitative variables. For PC dissemination, points were assigned to the six choices as follows: five points for "more than one unit per person," four points for "one unit per person," three points for "one unit per two to three people," two points for "one unit per four to five people," one point for "one unit per six or more people," and zero points for "no units available." For the adoption and use of ICT tools, there were three choices for each tool, namely "has been adopted and used" (two points), "considering adoption" (one point), and "no plan for adoption" (zero points).

Then, principal component analysis was performed on the basis of the points, and the first principal component with the highest eigenvalue was used as the index value. This is because, as shown in Table 2, the component weightings shown in the rotated component matrix indicate that each variable is making a substantial balanced contribution, and the first principal component represents comprehensive the ICT-intensive firms Finally, four specimen groups were formed using the principal component points to make the size of each specimen group almost equal. The group with the greatest progress in ICT use was labeled "level 1," and others were given descending scales to "level 4" for use in the analysis.

**Table 2.** Descriptive statistics of information and communication technology (ICT) measures

| Variable name | Mean | Standard deviation | N |
|---|---|---|---|
| ICT-intensive Level | 2.502 | 1.122 | 1431 |
| | | % | N |
| Level 1(with the greatest ICT-intensive) | | 25.2 | 360 |
| Level 2 | | 24.7 | 354 |
| Level 3 | | 24.9 | 356 |
| Level 4 (with the least ICT-intensive) | | 25.2 | 361 |

| Component | Total | %of Variance | Cumulative % |
|---|---|---|---|
| First | 2.413 | 34.474 | 34.474 |
| Second | 1.066 | 15.231 | 49.705 |

| Variable name used Principal Component Analysis | Mean | Standard deviation | Rotated matrix of the first principal component |
|---|---|---|---|
| Points for PC | 3.34 | .976 | .560 |
| Points for e-mail | 1.91 | .330 | .699 |
| Points for groupware | 1.57 | .666 | .650 |
| Points for Internet | 1.94 | .266 | .599 |
| Points for corporate website | 1.74 | .539 | .579 |
| Points for backbone system | 1.01 | .771 | .476 |
| Points for e-commerce | .88 | .702 | .518 |

### 3.3.2.3 Employment Arrangement in ICT-Intensive Firms

One-way analysis of variance was carried out to clarify whether there are differences in the labor force arrangement among the levels of ICT use. The percentages of temporary part-time workers, nonregular workers, and college graduates were used as dependent variables and the level of ICT use as the explanatory variable (Fig. 1). In the JIL Survey, the number of employees excluding temporary workers, the number of regular employees, the number of part-time workers, the number of temporary workers, and the ratio of college graduates to regular employees are known. Thus, the percentage of data that represent the boundaries in the labor market were obtained by making the denominator the "number of employees including temporary workers," for each denominator, namely the number of part-time workers, the number of temporary workers and the number of nonregular workers (the sum of part-time and temporary workers). For the percentages of college graduates, the results shown in the responses were used as received.

To clarify how the level of ICT use impacts on the future labor mix, correlations with the willingness toward future usage of external labor forces were surveyed. The question used for the analysis asked how ICT use would change the ratio of nonregular employees and the weighting of outsourcing among overall tasks in the future (about 3 years). The choices were "likely to increase," "will not change," "likely to decrease," and "will not be used." Fig. 1 shows the results of correlation analyses with sequential variables after deleting "will not be used."

Looking at the error bar in Fig. 1, the effects of ICT use on the current labor force organization can be grasped intuitively. First, attention should be paid to the "ratio of part-time workers" on the left. The ratio was the highest among organizations at level 4 in ICT use, and the ratio decreased with progress in ICT use. The average ratio for organizations with the highest use at level 1 was a little over 10%, and that for organizations with the least progress at level 4 was a little over 18%, showing a difference of about 8%. In short, the results are exactly the opposite of the common belief that the ratio of part-time workers increases as ICT use progresses. However, looking at the ratio of temporary workers shown second from left, we can observe that the ratio of temporary workers rose with progress in ICT use from 2.5% to 4.5% on average. In this respect, the common belief that the use of external labor forces progresses as ICT use progresses is supported. Looking at the percentage of college graduates, the hypothesis about SBTC seems to be supported.

How would these changes in the ratios of part-time workers as well as the changes in the ratio of temporary workers, change the ratio of nonregular workers, or the boundaries in the labor market? Looking at the error bar of the ratio of nonregular workers shown third from left (Fig. 1), the ratio is notably high at a little less than 21% on average among organizations that are level 4 in ICT use. The average for organizations at levels 3 to 1 is around 15%, with little difference among them. One feature of note is that the ratio of nonregular employees is slightly lower for organizations at level 1 in ICT use. In other words, the boundary between the external and internal labor markets has shifted in organizations at level 4, when compared with organizations of other levels. On the other hand, no shift in the boundary between labor markets due to ICT use was observed in organizations at levels from 3 to 1.

**Fig. 1.** Differences in employment arrangement according to information and communication technology (ICT) levels

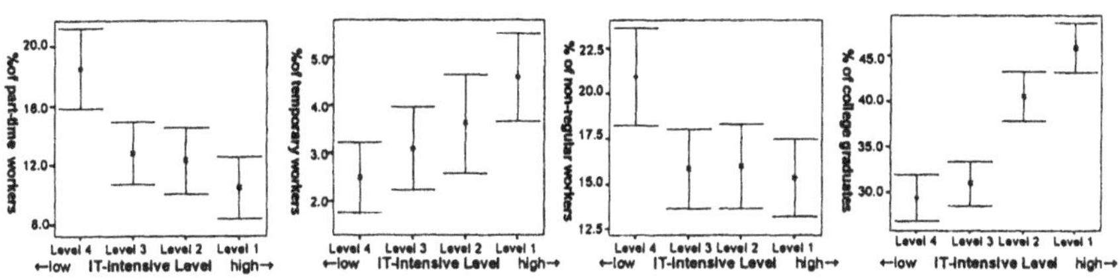


The *lines* above and below each error bar represent the upper and lower limits of the area of 95% reliability on average for each specimen group. The *dot* in the center represents the mean value for the specimen group.

It is possible that organizations at ICT level 4, where progress in ICT use is the least, the ratio of part-time workers could be high because it is difficult for them to use ICT to replace workers who take care of tasks that do not require experience. If the level of progress in ICT use were determined by the ease of ICT use for tasks, it could be possible to avoid a reorganization of the labor force in which the ratio of part-time workers decreases as ICT use progresses, while that of college graduates rises.

Even though there is no way to accurately study this point, the issue could be approached by surveying the levels of ICT use for different categories of business. Comparing the mean values of points for the main components of levels of progress in ICT use for each of the business categories with 10 or more samples, the level of ICT use was high among (1) information services (0.57), (2) electricity, gas, and water services (0.47), and (3) machine manufacturing (0.38), and was low among (1) transportation and communications services (−0.44), (2) financial and insurance services (−0.44), and (3) other services (−0.28). It is true that business categories with a low level of progress in ICT use include labor-intensive services. However, similarly labor-intensive wholesale, retail, and restaurant businesses (0.03) and foods (0.02) have levels of progress that rank them among the middle of all business categories. Therefore, it is not likely that ICT use has been delayed because digitization is difficult for tasks that do not require experience. Moreover, transportation and communications service providers and financial and insurance service providers at least partly belong to industries with high ICT use in the United States, and have improved productivity by using ICT (Dumagan et al. 2003). Therefore, ICT use does not seem to have been delayed because the effects of ICT on productivity are insignificant. As discussed below, sales have grown more among organizations that have promoted ICT use and have not relied on part-time workers, irrespective of the business category or corporate scale. Because it is essential for organizations to improve their business outcomes for them to survive, it is likely that even organizations that have not promoted ICT use to date will do so in the near future, and will rearrange their businesses, organizations, and labor force organizations to lower the ratio of part-time workers, in order to improve their business outcomes.

Next, we check to see whether there will be differences in the future labor force organization based on the ICT use levels (see Table 3). The correlation coefficient shows the statistical values by the Spearman test. When the correlation coefficient is positive with significance, more respondents expected that the willingness to use the variable would increase at organizations with a higher ICT use.

No statistically significant relationship was found between the willingness to use part-time workers and the levels of ICT use. The willingness to use temporary workers and outsourcing have statistically significant correlations with the levels of ICT use, and the correlation coefficient of outsourcing is greater than that of temporary workers. Firms expected the weight to increase more in the future for outsourcing than for temporary workers.

**Table 3.** Correlations between measures of ICT and future employment arrangements

| | Correlation coefficient | Mean | Standard deviation | *n* |
|---|---|---|---|---|
| Part-time workers | −0.004 | 1.38 | 0.598 | 1249 |
| Temporary workers | 0.089** | 1.34 | 0.602 | 1136 |
| Outsourcing | 0.175** | 1.51 | 0.562 | 1240 |

Correlations between the ICT use level and each of part-time workers, temporary workers, and outsourcing were determined by allocating two points if the weighting would "increase" in the future, one point if it would be "unchanged," zero if the weighting would "decrease," and by processing the responses "will not be used" and "no answer" with negative values

** Significant at $P = 0.01$

The use of external labor forces has also progressed at the manufacturing facilities that have supported the so-called Japanese-style employment practice. The survey on employment structures in the electric appliance industries in October 2000 declared the expanded use of contracted workers at manufacturing facilities (Satoh 2001). Follow-up surveys were conducted in November 2003 and April 2004.[10] Around the period in question, the Japanese economy experienced two bouts of recession with negative growth of gross domestic product. Business reorganization and employment adjustment were underway in the industry during the late 1990s to the early 2000s. The survey revealed that the number of regular employees and part-time workers fell as firms were restructuring with the adoption of the cell manufacturing system and other workplace reorganization, while the number of temporary workers increased significantly.

Dramatic changes in the employment arrangements from employment and training of full-time employees toward buying labor manpower and skills in the external labor market have not occurred. The results from analysis of firm-level data indicated exactly the opposite of the hypothesis that ICT use would cause regular employees to be replaced by part-time workers in the internal market,

[10] See homepage of Japanese Electrical, Electronics, and Information Union available at http://www.jeiu.or.jp/soken/index.html.

although the impact of a labor-saving effect does not appear in this section. Yet, highly ICT-intensive firms have increased the percentage of temporary workers instead of hiring more regular workers, and they are likely to do so in the future. While firms employ part-time workers mainly to reduce personnel costs, they hire temporary workers primarily to secure the requisite readiness and expertise. That is, the replacement of regular employees with temporary workers is done to improve productivity to pursue both functional flexibility and numerical flexibility. This approach poses more risk to outsiders and they need higher skill and must keep their skills up to date by continuous investment in self-training.

### 3.3.3 Changes in Wage System and Job Security of Core Workers

Figure 2 shows the results of the crossover tabulation regarding between the ICT use levels and the wage systems of the present and the future (about 3 years). More firms with higher ICT use levels responded that the present wage system is more inclined toward the merit system. Moreover, with respect to the wage systems of the future, no matter what the ICT use levels, only a few organizations responded that the seniority-based system would prevail, and more organizations with higher ICT use levels responded that the wage system would shift from that based on ability to that based on performance. The adoption of a performance-based wage system has statistically positive significant correlations with the level of ICT use at the 0.01 level. The correlation coefficient is 0.080 for the present and 0.200 for the future.

**Fig. 2.** Wage system becomes performance based as IT use progresses

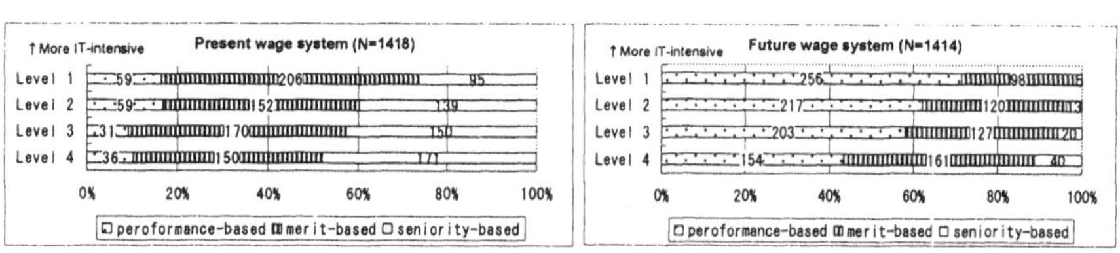

The performance-based wage system normally functions to transfer risks accompanying business fluctuations to employees in exchange for low levels of long-term commitment from employees. However, if the personnel appraisal systems are based on the relative levels among their employees rather than the absolute level of outcome, it does not necessarily result in wage and job arrangements that reflect the market, while it will lead to growth in within-group wage differentials.

Abe (2001) reveals that firms with high ICT use place emphasis on knowledge and skills obtained through work practices, rather than through general education, and also on the knowledge and skills obtained through on-the-job training, rather than off-the-job training. The skills may also be valid for the other firms because of flexible boundaries among firms, but they cannot be acquired without long-term training. Therefore, ICT-intensive firms encourage long-term commitment from employees and invest more than less ICT-intensive firms to construct an environment in which employees can bring out their potential and perform work creatively. We shall now look more closely at this point, using questions regarding the willingness among firms to support the skill development of their employees. Sequential variables were established for each of the items (1) through (7) shown in Table 4, assigning 2 to "already working on it," 1 to "wish to work on it in the future," and 0 to "not interested for the time being," and their correlation with the ICT use levels were surveyed.

**Table 4.** Measures taken by ICT-intensive firms to develop skills of their employees

| Variables | Correlation coefficient | Mean | Standard deviation | *n* |
|---|---|---|---|---|
| (1) Autonomous time management by employees by making working hours flexible | 0.286* | 0.89 | 0.789 | 1404 |
| (2) Establishment of an environment that enables concentration on work (e.g., reducing meeting times/hours) | 0.120* | 0.90 | 0.670 | 1395 |
| (3) Physical and mental reinvigoration through the adoption of special gift programs | 0.218* | 1.15 | 0.831 | 1404 |
| (4) Investment in the office environmentto secure independent working spaces | 0.187* | 0.61 | 0.664 | 1397 |
| (5) Improvement of measures for maintaining sound mental health to relieve stress | 0.219* | 0.79 | 0.668 | 1395 |
| (6) Revision of personnel appraisal systems so | 0.249* | 1.40 | 0.610 | 1408 |

| | | | | |
|---|---|---|---|---|
| that each employee's capabilities and achievements (contributions) are better reflected | | | | |
| (7) Promotion of information sharing to ease communication, coordination, and notification burden | 0.403* | 1.53 | 0.572 | 1417 |

The correlations between the ICT-intensive level and each sequential variables are the Spearman statistical values for those with significant Kendall's Tau-b
* Significance at P = 0.01

Looking at the results, the ICT use variable had a significantly positive correlation with variables on the investment with respect to the creation of the environment that allows employees to bring out their potential, at the 1% level, for all items (1) through (7) (see Table 4). In short, firms with higher ICT use levels have been active in developing the skills of the current employees. Reforms of conditions strengthen pressure to improve business outcomes on employees because firms with higher ICT use levels have supported measures such as (3) and (5).

Most firms appear willing to assure the long-term security of their employees. A total of 76.3% of respondents in the JIL Survey answered "take employment responsibility as possible" of labor demand changing driven by ICT-related technological progress. However, knowledge-based work systems can coexist with downsizing and slow wage growth as mentioned above. The ability of highly ICT-intensive firms to create new jobs is also important.

We use responses to questions regarding the change in sales and workers observed in the past 3 years in the JIL Survey. The choice for the answers to the questions are, "increased," "unchanged," "decreased," and "don't know." Because interest is primarily on the labor-saving effect, logistic regression analysis was performed using dummy variables as explained variables, namely 1, which was assigned to firms of which the sales/workers "increased" and 0, which was assigned to those of which the sales/workers were "unchanged" or "decreased."

Explanatory variables are the ICT use variable and four dummy variables of employment arrangement, controlling for industry (dummy variable of manufacture or nonmanufacture), the numbers of workers, and the scale of capital stock. Three dummy variables of employment arrangement are constructed: two groups for firms with the same specimen size based on the ratio of part-time workers, temporary workers, and university graduates; and the other one is based on whether the wage system based on the future performance is adopted. The results are shown in Table 5.

Regarding sales, level 2 and level 1 of ICT use are significantly positive and the coefficient for level 1 is greater than that for level 2. However, for workers, only level 2 of ICT use is significantly positive. There is no significant odds ratio difference for the growth of workers between level 1, the most ICT use, and level 4, the least ICT use. This result implies that the ICT use variable is correlated with the sales growth, but is not so with the workers growth. Examining variables of labor force arrangements, the larger percentage of temporary workers is significantly positive with the growth of sales and workers. Although, whether the growth of workers means the growth of regular employment or of both regular and nonregular workers is not clear in the survey. If changes in employment arrangements are taken into consideration, we cannot expect that full-time employees will increase as in the early 1990s.

**Table 5.** The acquisition of sales growth and employees for highly ICT-intensive firms

| Reference group | Explanatory variables | Sales | | Workers | |
|---|---|---|---|---|---|
| | | $\beta$ | Exp ($\beta$) | $\beta$ | Exp ($\beta$) |
| ICT use level | | | | | |
| | Level 1 | 0.550*** | (1.733) | 1.136 | (1.146) |
| | Level 2 | 0.437** | (1.548) | 1.424** | (0.527) |
| | Level 3 | 0.250 | (1.284) | −0.075 | (0.927) |
| Less ratio | | | | | |
| | Part-time workers | −0.004 | (1.004) | 0.094 | (0.098) |
| | Temporary workers | 0.318** | (1.374) | 0.255* | (0.291) |
| | College graduates | 0.207 | (1.230) | 0.274* | (0.315) |
| Seniority and merit based | Performance-based wage system | 0.087 | (1.091) | −0.085 | (0.919) |
| | Constant | −0.686*** | (0.504) | −1.513*** | (0.220) |
| | Likelihood | 1454.371 | | 1254.979 | |
| | Chi-square | 43.931*** | | 55.325*** | |
| | $n$ | 1111 | | 1121 | |

* Significant at $P = 0.1$, ** $P = 0.05$, *** $P = 0.01$

## 3.4 Conclusion

The recent changes in the employment environments of the United States and Japan are different. Wage differentials caused by skill premium for highly skilled workers and overall insecurities is increasing in the United States, while the diversification of employment patterns with discrimination against outsiders is occurring in Japan. However, both experiments driven by ICT-related technological progresses imply that we can no longer expect individual firms to accept responsibility for solving all labor-market problems.

The problem that must be resolved is how to link productivity growth with prosperity growth. It is partly a skill problem, because recent technological changes favor highly skilled workers. The knowledge-based work system should be encouraged, but it does not promise mutual gains for overall employment. It is certainly important to promote a relative decrease in the wage of the skilled and to increase training in order to shift a portion of the unskilled into the skilled pool by facilitating innovations of private sectors and constructing the cross-group coalitions for enhancing long-term employability.

It seems that the absolute demand shift is more important than the relative demand shift. If labor-saving technological change reduces the absolute number of workers, both the skilled and unskilled workforce will be displaced. This means that the supply-side approach to resolving a skill problem can deal with only limited problems. In order to resolve overall labor market problems, the solutions for the demand side should also be emphasized.

## Acknowledgments

The author expresses sincere gratitude to the Social Science Japan Data Archive at the University of Tokyo for providing a data set of the JIL Survey. The author is also extremely grateful to Dr. Sudoh (Professor, The University of Tokyo) and an anonymous referee for providing essential comments. The work described in this chapter has been supported in part by a Grant-in-Aid for one of the Specified Research Areas, "Informatics Studies for the Foundation of IT Evolution," which is founded by the Ministry of Education, Culture, Sports, Science and Technology.

## References

Abe M (2001) How are information and communication technologies affecting employment? (in Japanese). Japanese Journal of Labour Studies 498:13–26

Alchian AA, Demsetz H (1972) Production, information costs, and economic organization. American Economic Review 62:777–795

Anon. (2001) IT adoption and organizations/labour (in Japanese). The Japan Institute of Labour, Tokyo

Autor DH, Katz LF, Krueger AB (1998) Computing inequality: have computers changed the labor market? Quarterly Journal of Economics 113:1169–1213

Bartel AP and Sicherman N (1999) Technological change and wages: an inter-industry analysis. Journal of Political Economy 107(2):285–325

Berman E, Bound J, Machin S (1998) Implications of skill-biased technological change: international evidence. Quarterly Journal of Economics 113(4):1245–1279

Bresnahan TF, Brynjolfsson E, Hitt LM (2002) Information technology, workplace organization, and the demand for skilled labor: firm-level evidence. Quarterly Journal of Economics 117:339–376

Brynjolfsson E, Hitt LM, Yang S (2002) Intangible assets: computers and organizational capital. Brookings Papers on Economic Activity 1:137–181

Cappelli P (1999) The new deal at work: managing the market-driven workforce. Harvard Business School Press, Boston

Christensen CM (1997) The innovator's dilemma: when new technologies cause great firms to fail. Harvard Business School Press, Boston

Cooke SD (2003) Information technology workers in the digital economy, in U.S. Department of Commerce. Digital Economy 2003, Washington DC, pp 19–34

Doeringer P, Piore M (1971) Internal labor markets and manpower analysis. Heath, Lexington, MA

Dumagan J, Gill G, Ingram C (2003) Industry-level effects of information technology use on overall productivity, in U.S. Department of Commerce. Digital Economy 2003, Washington DC, pp 45–60

Economic Planning Agency (2000) Productivity effect of IT (in Japanese). Policy Effectiveness Analysis Report 4

Genda Y (2001) Ambiguous insecurity in work (in Japanese). Chuo Koronsha, Tokyo

Gordon DM (1996) Fat and mean: the corporate squeeze of working Americans and the myth of managerial "downsizing." Free, New York

Gotoh R (2004) Employment arrangement in the information society (in Japanese). In: Sudoh O (ed) Proceedings of International Conference: Establishment of Information Economy and Evolution of Socioeconomic Systems Tokyo, pp 87–101

Groshen E, Potter S (2003) Has structural change contributed to a jobless recovery? Current Issues in Economics and Finance, Federal Reserve Bank of New York 9:2–4

Haltiwanger J, Jarmin R, Schank T (2003) Productivity, investment in ICT and market experimentation: micro evidence from Germany and the U.S. Center for Economic Studies Working Paper CES-03-06, US Bureau of the Census

Hand J, Lev B (2003) Intangible assets: values, measures, and risks. Oxford University Press, Oxford and New York

Harashima T, Tejima H (2002) Actual wage gaps (in Japanese). In: Miyajima H, Rengo Research Institute for Advancement of Living Standards (eds) Income distribution and gaps in Japan. Toyo Keizai Shimpo-sha, Tokyo, pp 49–98

Jorgenson DW, Motohashi K (2004) Potential growth of the Japanese and U.S. economies in the information age. ESRI Discussion Paper Series 88:1–11

Kaplan RS, Norton DP (2004) Strategy maps: converting intangible assets into tangible outcomes. Harvard Business School Press, Boston

Lipietz A (1992) Towards a new economic order: postfordism, ecology and democracy Europe and the international order. Polity Press, Cambridge

Machin S, Van Reenen J (1998) Technology and changes in skill structure: evidence from seven OECD countries. Quarterly Journal of Economics 113(4):1215–1244

Mishel L, Bernstein J, Allegretto S (2004) The state of working America, 2003–2004. M.E. Sharpe, Armonk, NY

Nicoletti G, Scarpetta S (2003) Regulation, productivity and growth: OECD evidence. Economic Policy 36:9–72

Ohara M, Otake F (2001) The impact on wages of computer use (in Japanese). Japanese Journal of Labour Studies 494:16–30

Ohsawa M (2002) Polarization in the labor market brought about by increase of non-regular employees (in Japanese). In: Miyajima H, Rengo Research Institute for the Advancement of Living Standards (eds) Income distribution and gaps in Japan. Toyokeizai Shinbunsha, Tokyo, pp 119–151

O'Reilly CA III, Pfeffer J (2000) Hidden value: how great companies achieve extraordinary results with ordinary people. Harvard Business School Press, Boston

Osterman P (1999) Securing prosperity: the American labor market: how it has changed and what to do about it. Princeton University Press, Princeton

Osterman P, Kochan TA, Loche R, Piore MJ (2001) Working in America, MIT Press, Cambridge, London

Piore MJ, Sabel CF (1984) The second industrial divide. Basic Books, New York

Satoh H (2001) Employment systems in the IT age (in Japanese). Nippon-Hyoron-Sha, Tokyo

Scott Morton MS (1991) The corporation of the 1990s: information technology and organizational transformation. Oxford University Press, New York, Oxford

Takayama Y (2004) Nonstandard work in the U.S. (in Japanese). Japanese Journal of Labour Studies 524:33–44

Vivarelli M, Pianta M (2000) The employment impact of innovation. Routledge, London

# Part II: The Nature of Competition

# 4 Organization–Issued Cash in a Digital Economy[1]

Hiroshi Deguchi

Interdisciplinary Graduate School of Science and Engineering, Tokyo Institute of Technology, 4259 Nagatsuta-cho, Midori-ku, Yokohama 226-8502, Japan, deguchi@dis.titech.ac.jp

**Summary**

In this chapter we focus on the possibility of cash issued by an organization as in the form of micro-bonds that are redeemed with the added value created by the organizations. We investigate several scenarios where a platform of organization-issued cash (OIC) could grow. We insist that an age of ultrapluralistic money based on pluralistic values is possible for all kinds of organizations that form added value and issue OIC, but it is unthinkable without IT.

**Key Words** Organization-Issued Cash, Network valuation landscape, Community money, Electronic public sphere, Pluralistic valuation

[1] This chapter is based on research that was conducted at the NTT Open Laboratory and was reported in November 1999. (NTT Open Laboratory Symposium, Social Systems Promoting the Distribution and Sharing of Information, First Term Open Laboratory Activities Final Report Meeting, "Economic Transactions Research Group Report." 19 November 1999, NTT Musashino Research and Development Center (see http://www.openlab.tnl.ntt.co.jp/event/symp5/Sudoh/Sudoh.html).

## 4.1 Introduction

It is already becoming abundantly clear that a new global industrial revolution, or more correctly an information network revolution, is taking place within our economic systems. Network reforms of business models and processing changes are occurring one after another; forcing even greater reforms on existing corporate organizations and their industrial structures. These days it is difficult to argue with any conviction just how long these reforms and trends will continue or ultimately what the shape of a reconstructed global economic system will be (Deguchi 1994, 2004).

Without any doubt, one of the greatest changes that could occur in the future digital economy is the reform of systems of valuation, i.e., the issue and circulation of credit. Such reforms of electronic credit will fundamentally change the nature of the broader socioeconomic systems, along with the activities of many organizations, including the companies that operate such systems.

During this process, there is the possibility that the conventional nature of money and the dynamism of valuation and choice in a capitalist economy will change; in a capitalist economy, the dynamism of capital markets in relation to investment lies at the root of capitalist competition. Channels have been created that enable organizations to add value and then link the future ability to add value with the current credit supply. Companies that produce goods or services are evaluated based on the purchasing behavior of consumers in the market, and in the capital markets by the valuations of speculators. As a result, it is possible in a competitive market environment for corporations to design and deliver goods or services that are likely to be met with market approval.

It is precisely within this mechanism of providing capital, through the valuation of organizations with the ability to create added value in the form of the stock market, that the dynamic heart of capitalism lies, and, as a result, a social circuit of credit supply is created by speculators in such a way that the production of future added value is linked to current capital investment. However, this circuit does not always function effectively for those organizations that do not aim to maximize profit in relation to capital. There is no guarantee that adequate credit creation and supply will be afforded for the activities of nonprofit organizations (NPOs) that create socially useful added value. Such organizations do not seek profit maximization, do not provide services in the form of public goods or common goods, or engage in activities designed to target diverse markets. On the contrary, they are often very individual in character, operate on a small scale, and generate a small profit. To some extent, the twentieth century type of national economic system, supported by national government, has helped to balance these inequalities; however, it is difficult for this system to obtain the dynamism of effective valuation and choice. Many public corporations and third sector enterprises still have problems, and there is no guarantee that effective competitive resource distribution is carried out in Japan. In the meantime, today's capital markets have also become more complex with a variety of derivatives, and as a result have come to experience instability unconnected with their original function of credit creation and supply.

There is a need to design a social circuit to underpin the infrastructure of the new network society; a mechanism that evaluates the added value of the goods or services produced by the various bodies in society. This would include profit-making organizations such as companies, and midway organizations such as social networks, as well as the various nonprofit corporations that also create and supply credit. To this end, this chapter examines the possibility of organization-issued cash in the form of micro-bonds that are redeemed with the added value created by the organizations.

This chapter analyzes the significance that credit creation and supply by electronic organization-issued cash would have for the various operational bodies in society, including not only companies but also NPOs. It also explores scenarios for launching such a venture, and how electronic coupons, points, and community money, as examples of electronic money, would converge into the organization-issued cash proposed by this chapter. New business models are also proposed for using organization-issued cash. Then, from two different viewpoints, i.e., high-economic private money theory where money is "the most marketable commodity" and Carl Menger's money commodity theory (Menger 1984), this chapter sketches a grand design for a new age of electronic private credit creation by organization-issued cash, which allows pluralistic valuation and choice, and credit creation and supply.

## 4.2 Reform of the Credit System and Money

As the digital economy has developed there has been much speculation concerning the new possibilities of electronic money. Unfortunately, many of these concepts have been rooted firmly in a classical interpretation of money; a view that considers electronic money as "a means of electronically expressing information backed by actual currency" (Iwai 1998). This is little more than an interpretation reflecting our preconceptions of what we call "money." The question of how money came to win confidence is the most interesting process in the history of money. Historically speaking there was a time when the legal tender of paper money and hansatsu (paper money issued by the feudal clans of the Edo era whose currency was limited to within clan boundaries) was also considered as little more than information backed by gold or other "goods believed to actually have value."

Let us assume we accept the general view that the basic function of money is the expression, saving, and exchange of value. This expression of value by money is the flipside of the valuation of money itself, that is, how does money acquire its own value. The view of electronic money as "a means of electronically expressing mere information backed by actual currency" or as "a liability to be redeemed with something that actually has value" is the view that it is state-issued currency with a truly guaranteed value and electronic money is then little more than information that guarantees exchange. Current models of electronic money issue, especially closed loop models, are indeed based on this kind of conceptual framework. However, if we consider how, in historical terms, currency in the form of convertible paper money has shifted to unconvertible paper money, there are clear difficulties with a view that places the source of confidence solely in the currency already in circulation. The Japanese hansatsu of the Edo period were not a very reliable credit medium, and, as in many regions and periods, precious metals such as gold and silver were the preferred means of expression, saving, and exchange of value. Convertible paper money, which in modern times has come to be issued by banks in many countries, has value because it could be exchanged with the original means of expression, saving, and exchange of value: gold. The basic principle of convertible currency sees it as something that realizes the function of a paper money currency that secures confidence by invoking a means of saving and evaluation of value, that is believed to be stable, i.e., gold (exchange a note for gold). Before long this develops into an unconvertible currency system of national units based on confidence in government and this in itself then comes to earn a confidence all of its own. In other words, currency in the form of convertible paper money was little more than "a means of symbolically expressing mere information backed by gold, which is believed to actually have value" or "a liability to be redeemed with something that actually has value." This grew into unconvertible paper money because nation states—as the territorial governing bodies encompassing many organizations including corporations capable of added value production—and their central banks came to win sufficient confidence. In this process, the privately issued bank notes of commercial banks were absorbed and integrated into the system of central bank notes.

Historically speaking, confidence in money as legal tender has shifted from confidence in the nation that guarantees the exchange of convertible notes for gold, into confidence in the nation that issues the currency and manages the economy of the region of its circulation. This process was completed with the nonconversion policy of the American dollar. The history of inconvertible paper money was also the history of how nation states have become strong. Conversely, even today the currency of a country, where the added-value production of the economy within that region is unstable, is constantly exposed to a crisis of confidence regardless of the size of the country.

A major role of such currency is to supply credit to organizations and individuals that have the potential to generate added value in the future. An economic system only gains really great dynamism through such a credit supply. The question of how to create this as a system that functions socially was an important problem for economic systems. This entails the two issues of maintaining confidence in the currency itself and supplying credit to organizations and individuals that need currency. These two issues are closely connected, but their effect on the system is different. To maintain confidence in the currency, itself a mechanism that evaluates and chooses the currency, is the issue. A valuation of currency can be made in reverse, based on the goods produced by society, for example, in the form of the inflation rate. A currency can also be valuated and chosen by comparison with other currencies. Of these two types of valuation, it is a fundamental function of legal tender that the reverse-valuation by goods is stable. For the latter credit supply, the design of the system for valuation and choice, in relation to how organizations and individuals are evaluated and supplied with credit, is the issue. The first aims to maintain confidence; it is considered necessary to evaluate the future ability of organizations to create added value according to purchasing behavior on the market of products or services on a different level from the valuation made of the companies that produce them. The second aims to supply credit to promising companies by evaluating in advance the profit that could be obtained over time as a result of corporate capital investment. To do this there is the mechanism of valuation and choice for indirect finance by financial organizations and valuation and choice on capital markets relating to the bonds and stocks issued by the companies themselves. In particular, the system of the stock market is a brilliant invention of capitalism that evaluates corporate activities that give rise to future added value on the basis of the maximization of profit or capital gain. It creates and supplies credit that can be secured through speculation on the instability of capital investment and does not need to be repaid.

In contrast, local credit creation and supply by the private sector such as corporate bills and bonds is possible because a company guarantees that it will repay the liability in currency and has sufficient confidence with respect to the repayment. This assumes that the bills and bonds will be recovered and repaid in currency, and this in itself is little more than a limited means of credit creation and supply and the function of expression, saving, and exchange of value is limited. This does not mean to say that with this in itself some kinds of goods are evaluated. In other words, it is not a "liability to be repaid with something that actually has value," rather it is "a liability that will be repaid with currency." In this sense, this does not have a money function.

What kinds of new possibilities would the electronic expression, saving, and exchange of value or creation and supply of credit, such as the electronic money now being developed, have through Internet information technology? Within financial organizations, offsetting transactions are conducted daily using electronic settlement systems, and, in recent years, the settlement of mutual offsetting transactions called "netting" has become common between corporate groups. This is an electronic mechanism that not only reduces exchange risks and charges, but also internally circulates local credit in the form of credits and obligations issued by the companies themselves without exchanging them for actual currency, but uses the existing currency expression function. In such a case, if liabilities are repaid with something that has value independently of legal tender, this means that money in the form of "a liability that is repaid with something that actually has value" is created.

## 4.3 Organization–Issued Cash

### 4.3.1 Local Money Issued by Organizations

We need a system in the form of a new social circuit that enables us to create and supply credit for activities such as capital investment for future value creation of value, carried out by organizations with a valuation basis other than profit maximization. One possibility that links multidimensional network evaluation with resource distribution is organization-issued cash in the form of micro-bonds or coupons issued in such a way that they are repaid with the added value formed by the organizations themselves. Here open-loop electronic money, in the form of micro-bonds that are issued by organizations and repaid with the added value created by the organizations themselves, functions within an electronic market where they are mutually evaluated and exchanged. Such an entity is called organization-issued cash (OIC) in this chapter and of course, in reality such cash does not yet exist. Let us discuss how cash in the form of micro-bonds issued by organizations and redeemed with added value created by the organizations themselves is possible and in what sense this cash is essential to a digital economy.

In the case of companies, for OIC to be possible, it is necessary for the products supplied to be valued using company OIC. Let us assume that companies issue their own OIC units to value the goods or services that are the added value they produce or provide; and such OIC units are called company-issued cash (CIC) in this chapter. For example, let us assume Sony Corporation issued its own company cash called So_cash and then valued and priced its own products using its own cash unit called a Sonit. Based on this assumption, a customer with access to So_cash would be able to purchase all the products or services of Sony Corporation using So_cash. Let us also imagine that Microsoft Corporation issued Micro_cash and priced its own software in units called Minit or Gatu, thus making purchases in Micro_cash possible. Then the services of affiliated companies such as the computers of PC makers whose operating systems (OS) are supplied by Microsoft Corporation could be priced and purchased in Micro_cash. In that event, because these companies purchase OS from Microsoft Corporation, this would open the door to pay for the OS in Micro_cash. In this way, provided it is used within an open loop, electronic money in the form of company-issued cash and issued by one company will be acceptable and free to circulate inside another company.

In relation to electronic credit, such company-issued cash offers only one possible business model; electronic money is versatile and can represent money as well as coupons, points, or prepaid cards, all of which can offer incentives—discount services and special offers—to reward, retain, or gain customers. This kind of company-issued electronic money can also be used for settlement between corporate groups and in that case, dynamism arises that is clearly different from credit supply as the means of fundraising of corporate bonds and that in itself earns value. What endorses that value is the added value produced by the company, in other words, a valuation of the products or services produced by the company. To take an extreme example, if there were premium goods that could only be obtained with CIC issued by a company, confidence in the CIC in question would probably rise depending on the popularity of the premium goods. In other words, the CIC would win confidence depending on the added value that is the products or services offered by the company.

The issuing bodies of such OIC are not confined to profit-making companies; provided they are bodies that can create some kind of added value, NPOs and nongovernment organizations (NGOs) can issue their own cash using the same mechanism. In that sense, we use the term organization-issued cash or OIC to include both company-issued and NPO-issued cash. OIC expresses the added value of the goods or services an organization produces, saves that added value, and also has a local exchange function. In addition, confidence in the OIC depends on the added value production ability, i.e., the goods or services of the organization, and people's evaluation; valuation is not only carried out by the markets. Today, in relation to various social problems, nonmarket valuation and valuation networks are being formed on information networks, and various attempts have also been made to link such nonmarket valuation and market valuation (Tanimoto 2000). One such example in which OIC is opening up new possibilities is funds, such as green funds and ecofunds. People can buy and circulate OIC in the form of donations to support the activities of organizations that have the characteristics of public goods and are not purchased and consumed by individuals directly, such as volunteer activities and environment protection activities (service-added value production). As explained later, this would no doubt function effectively if it were combined with a tax system that allowed people to choose to donate part of a tax.

The value of OIC changes depends in a broad sense on the reputation of organizations beyond the capital markets. Therefore, the valuation of the OIC issued by countless organizations changes dynamically in the form of mutual exchange rates or against a base currency. Their valuation is reflected in the reputation of an organization and the possibility that such a dynamic form as the reputation of an organization can be assessed beyond the capital markets is one important possibility of OIC. Already companies in the market face three valuations: by banks for indirect finance, by CBs for direct finance, and by the market in the form of the stock price. However, there have always been doubts over valuation for indirect finance. Moreover, a shift is occurring from indirect finance — where financial organizations supply credit following the valuation of an organization — to direct finance, where confidence in an organization is evaluated and considered to be secured by the market. However, the basis for this type of valuation is purely that of stockholder profit maximization. Methods for adjusting the disparity between this basis of valuation on capital markets and a broader social evaluation include adjustments through government subsidy, fines, and partial use of the markets by ecofunds, etc. Alternatively, there is also the possibility of using indirect finance to give preferential treatment for loans to minorities and ecoloans etc. However, there is no guarantee that such mechanisms will correctly reflect dynamic changes in social valuation. Assuming an information network environment, it is necessary for us to free ourselves of the paradigm of government compensation for market failures and redesign a mechanism of resource distribution based on pluralistic valuation and competition. This is why we are focusing on the possibility of OIC as one avenue for dynamically forming direct valuations of the added value produced by various organizations, in a way that includes markets and various networks. The introduction of transferable and exchangeable OIC would enable a more flexible valuation of the ability to create added value and the creation and supply of credit.

### 4.3.2 Formation of Valuation Landscapes and Network Valuation

Credit supply to companies by direct finance is conventionally carried out on the basis of valuation and choice by the markets. However, indirect finance by financial organizations also plays an important role in supplying credit to companies. In an economy where it is difficult to take on adequate risk through direct finance or where strong investor groups have failed to emerge, indirect finance has played an important role. Previously, in Japan it has been believed that the banks responsible for indirect finance have both the ability to evaluate and to develop business models for the supply of credit. However, nowadays these abilities and the ability to supply credit quickly are in doubt and attempts are being made to create a new framework for credit creation and supply, such as venture capital.

However, with a credit supply mechanism that purely aims for maximization of investment return, it is impossible to afford adequate credit supply to bodies creating added value other than profit-making companies. The corporate valuation of banks and investors aims to maximize profit, in the short term or the long term. However, it is not socially meaningful to supply credit solely to companies that post profits and continue to grow. The Internet technology that has evolved on information networks has developed along the very different lines of convenience for software engineers and users and maximizing the development speed and not the maximization of profit. This is because various network resources in the form of volunteers, for example, have been supplied in response to the network valuation. Furthermore, with the open source software that has developed on networks, there is not necessarily only one basis of valuation. For example, in the case of the Unix system software Linux and BSD, groups have formed according to their respective and different sense of values (valuation landscapes), and, while competing with each other over their valuation landscape, have gathered supporters, developed and technically evolved. Furthermore, neither of them has been evaluated from the perspective of profit maximization. However, as a result they have accumulated capital goods on the Internet in the form of vast free software that is not counted in the ordinary GDP. This has brought enormous advantage not only to ordinary users but also to the software industries of developing countries.

If such network valuations were created socially and circulated as information, the method whereby only speculation and product purchases through markets establish the valuation landscape of goods and services would disappear. In the case of nonprofit organizations, the aim of such organizations is not just to grow.

Moreover, even if there were enough demand for the creation of added value such as services in areas where it is hard to expect high profit, or services in the form of public goods, it would be difficult to develop profit-making companies to supply them. On the other hand, there is also no guarantee that a mechanism for proper management evaluation designed to ensure the continuation of business and credit creation and supply could provide to the nonprofit organizations active in these areas. A framework of adequate evaluation and competition is also necessary for such nonprofit organizations and today such frameworks for NPO valuation are starting to develop, along with attempts at financial support for NPOs by banks, such as community banks. However, if there was a decentralized system enabling the creation and supply of credit that reflected the network valuation of the goods or services themselves created as added value, this would no doubt have a unique role not found in the markets. This is why there is a need within society for a stable form of network valuations and the creation of a system to enable credit creation and supply in accordance with such network valuation. OIC is certainly a system with the potential for constructing a careful and robust framework for the creation and supply of local credit, but this potential will only be realized through the application of electronics and networks.

## 4.4 OIC and the Local Cash Function

Let us describe any system with some kind of credit creation function as a system with a partial cash function. For example, the coupons and points issued by organizations and bodies; local money in the form of regional currencies, CBs, ecomoney, and stocks are all systems with a partial cash function, in the sense that they lack one of the various functions of currency.

Let us then describe local cash as the part of such a system whose function is to evaluate economic added value, i.e., some kind of goods or services, and to exchange with goods or services. In the above scenario, coupons, point systems, ecomoney, and local money are all systems with a local cash function. When discussing the possibilities of OIC, along with the possibilities offered by such forms of cash, the analysis of such systems with only the partial ability to express the saving and exchange of value and the issue and circulation of credit is essential.

### 4.4.1 Point Systems and Company–Issued Cash

In the actual economy, systems similar to OIC are now being used, albeit partially and for a variety of different aims. A typical example is the point system; point systems are used by many companies to win the exclusive loyalty of their customers. Under such systems, points are used as a kind of discount system, but interestingly at the same time, product pricing in points has also begun. For example, product prices are clearly shown using points like Softmap Corporation's "rupees" (see http://www.sofmap.com/). Point systems are restrictive and generally focus on winning and keeping customers by being nontransferable. The point-processing information systems are also linked with customer management systems using POS and such systems are not designed to enable the transfer of points between customers using an open loop. However, if they were circulated in an open loop, theoretically points would be very similar to OIC. Currently business models related to so-called point systems are being developed in a manner that links them with electronic money.

### 4.4.2 Coupons

Coupons in this context denote systems that have their own unit of value and when saved can be exchanged for designated goods. Coupons are sometimes issued incidentally as a kind of service when users have purchased designated products and services and are sometimes distributed to the general public as an incentive or discount on the purchase of services and goods. Coupons are sometimes made nontransferable offering a particular service to a named user and sometimes they are not. Depending on the way they are issued, coupons are no different from point systems. In terms of business models, like point systems, coupons that are issued to win customers and are accumulated by customers are nontransferable. However, coupons in a form that are in fact transferable and can be accumulated, such as the airmiles of airline companies, are starting to increase. In addition, various forums for the trade and transfer of such rewards have emerged on the World Wide Web. The issue of such coupons is also essentially the issue of a form of credit, as the coupons are backed with the products issued by the company. If an airline company reduces the number of flights, the mileage exchange value falls. In recent years, airline companies have extended the validity period of mileage-type coupons because when validity periods are short, high numbers of redemptions in exchange for air tickets has a negative effect on revenues and profits. In addition, recently, mileage points themselves have even been sold for cash as a service for people whose mileage points have nearly reached the required level, but more interestingly it demonstrates that the points themselves already have a certain exchange rate against cash.

### 4.4.3 Prepaid Cards and Gift Tokens

As shown with telephone cards, prepaid cards are a system that provides a more convenient or cheaper service after prepayment than can be provided when cash is used. Such prepaid cards and gift tokens can of course be considered as a type of limited OIC. They also supply credit in advance for the goods or services provided by the company. Advance movie ticket sales are a typical example. Tickets sold in advance could also become worthless if, for instance, the movie was not completed or for some reason was cancelled, but generally they are issued before the completion of a movie. Furthermore, prepaid cards are also usually transferable and this is why they are easily convertible into cash. In particular, so-called gift tokens such as beer tokens and department store gift tokens are handled in the same way as monetary notes, have a high liquidity at monetary note shops, and are exchanged with each other. In recent years, even telephone cards can be exchanged more widely, for example, they can now be used to pay public charges. In general, the issue of this kind of prepaid card and gift token clearly has the financial advantage of credit creation and supply for the issuing organization.

Seen in this way, it is clear that the boundaries between coupons and points, prepaid cards and monetary notes are unexpectedly low. If anything, the problem now lies in the creation of a system for open loop circulation and free exchange and in the design of business models that remove the barriers between these diverse systems of local cash. In certain areas, attempts at creating business models that remove these boundaries are beginning to be made.

Such schemes as airmiles and other point and coupon services are now a common part of the marketing and sale of goods and services and the eternal endeavor to win customers by offering incentives. As explained earlier, this type of service is not generally based on transferability, but there is the possibility that once transferability begins to spread as a service, systems would transmute rapidly to enable mutual exchange.

### 4.4.4 Community Money

There are various types of community money including the Local Exchange Trading System (LETS) popular in Canada and the United Kingdom, the American Time Dollar and Ithaca Hours, and the Swiss WIR (Wirtschaftsring-Genossenschaft) (Solomon 1996). However, they can be divided broadly into money that was created in opposition to a tightening of credit in the regional economy: the WIR was a response to the credit squeeze at the time of the great depression in 1929, and more recently, money with a strong base in community ties, such as LETS and Ithaca Hours (see http://www.wirbank.ch/ and http://www.lightlink.com/hours/ithacahours/).

Such community-oriented money is issued with several aims. Historically, legal tender could not cover a region's credit creation on its own and community money was conceived as a means of making up for a region's diminishing ability to create added value, due to a shortfall of credit. It also helped to prevent the outflow of added value created in the region. It is a mechanism whereby members of a community use the same money to create and circulate credit amongst themselves. For example, in the case of the American Ithaca Hours currency, 1 hour corresponds to US$10, converted into one hour's work. This can also be described as a type of coupon linked with the added value of labor. The linkage of this type of money with information technology makes it possible for a completely new type of OIC to emerge. In the case of the Japanese ecomoney system advocated by Kato (2001), valuation of care and environmental activities are also included, but its properties as community money are the same (see http://www.ecommunity.or.jp/). The characteristics of such community money include the following:

1. Community money interest is negative or zero. This effectively increases the circulation speed of the money and is effective in securing a reduction in the burden of risk to a business with credit within the community.
2. Community money may be issued in a form that is measured according to labor hours or mixes legal tender according to an agreed ratio.
3. The circulation of community money is limited to the area of the community.
4. The scope of products and services that can be exchanged with community money is limited.

However, this does not mean to say that all community money has the same characteristics. There are also various methods of delivering these characteristics, such as recording on a central server, points, entry in the form of a passbook, or the issue of paper notes expressing a value such as coupons.

For example, Swiss WIR is the most prominent business-originated community money and is managed centrally. In contrast, the American Ithaca Hours and Time Dollar or the Canadian LETS are community monies that stress community ties. In Japan also, various community money experiments are being carried out such as Peanuts and Ohmi (see http://www.seaple-.icc.ne.jp/murayama/letspeanuts.html and http://ww.kusatsu.or.jp/ohmi/).

The reason community money is considered separately from the currency of the country is that the system of credit creation and supply by the actual currency has a problem. The problem lies in the terms of its function of evaluating the hidden potential of organizations and individuals to create added value and its ability to supply credit to profit-making or nonprofit added value-creating businesses in that region. Even if the diverse bodies operating in society are engaged in activities that create added value and contribute to society in the form of goods or services, there is no guarantee that they will be afforded adequate credit valuation and credit supply under the market mechanisms that carry out indirect and direct finance.

A mechanism is needed that is different from the credit supplied by finance and speculation aimed at profit gains, or the compensation support given by governments. The mechanism must evaluate both the potential of a region or organization and its nonmarket added value-creation ability and then create and supply credit. The idea of OIC has wide possibilities as a system for various types of OIC, including such community money, to form networks amongst each other. We foresee a world where supranational medical group coupons, NASA fly-a-lander-to-Jupiter coupons, environmental protection activity coupons, and business coupons issued by companies are intermingled, instantaneously exchanged, and circulated electronically. Such examples of OIC are each issued and redeemed by different methods, but can be exchanged with each other. For example, supranational medical group bonds would be purchased under the donation tax system described later, subjected to a business evaluation some time later, and redeemed as a donation provided the buyer judges that satisfactory results have been obtained. NASA fly-a-lander-to-Jupiter coupons would be purchased in exchange for the right to be able to quickly obtain information about Jupiter or the right to have an engraved name plate on the lander. Environmental protection activity coupons would be purchased as a pure donation by the individual, subjected to a business evaluation some time later, and redeemed as a donation provided the buyer judged that satisfactory results had been achieved. Business coupons are redeemed against products or services provided by the company; as explained later, business coupons open the way to a variety of issuing business models.

## 4.5 For OIC to Become Established

### 4.5.1 Platform of OIC

Bonds are corporate liabilities and as they are in circulation, they are able to partly play the role of a means of settlement. When legal tender was convertible currency, paper notes were convertible notes and were a form of no-interest liability. Legal tender was guaranteed by gold held by the state and at the same time confidence in the legal tender was backed by the economic power of the state. In contrast, the bonds issued by companies are not convertible but their value is guaranteed by the fact that they will ultimately be redeemed against legal tender. Of course, bonds are issued as a means of direct finance, and for those who buy them it is a question not only of confidence in the bonds and their liquidity but also their profitability. However, provided their redemption is guaranteed, aside from the risk relating to profitability, they can have the attributes of currency as a means of settlement.

If an adequate electronic market were established, like OIC, micro-bonds redeemable in the form of added value, i.e. the goods or services provided by an organization, would also have the attributes of currency. Such micro-bonds would also be a very attractive product for speculators provided proper profit risk management was possible. On the other hand, under diverse evaluation landscapes, the circulation of OIC issued by organizations that do not aim for profit maximization could also be expected.

Micro-bond type OIC could be given the function of credit creation and supply through the imposition of a redemption time delay. It may seem that such micro-bonds are lacking as a means of long-term stable credit supply and are also small in scale, but this is probably due to their product design as a financial product of OIC.

Micro-bond type CIC could be purchased by investors and widely sold and circulated to general consumers. For consumers, this means the possibility of being able to purchase those products and services at low cost. A type that could be redeemed for cash a few years later, but could be also be redeemed for products immediately would be possible. Financial product design for various types of OIC would be possible.

If OIC with the function of local credit creation such as electronic coupons were used widely, then a crossover with various types of bonds would occur. When used as incentives, the boundaries of electronic coupons and point systems and bonds issued and circulated in micro-issues would also be able to cross over. Whereas the bonds are redeemed in cash, in principle OIC is redeemed in the added value produced by that organization. Micro-bonds, which are thus issued by organizations and redeemed with added value in the shape of the services or products produced by that organization, are similar to prepaid cards in one respect. If anything, coupons are understood to be a marketing tool that provides an incentive to purchase goods or services. However, there is the possibility that a model for direct finance using various types of micro-bonds could be formed at the intersection of coupons as a means of payment, electronic money for check-type payment, and micro-bonds redeemed with the added value issued by organizations. This is a model whereby bodies with the ability to create added value carry out credit creation themselves by issuing and circulating OIC using electronic micro-bonds. Moreover, it is a model whereby OIC becomes the object of speculation and at the same time serves as an exchange medium and means of marketing. Under this system, there is a high level of incentive for general consumers to obtain OIC that can be exchanged with designated products to pay for those products on purchase.

In this chapter, the claim is made that there is the possibility of a new credit supply and exchange and for the construction of marketing models in the form of a mixture of electronic bond markets as a means of credit supply, electronic money as a means of electronic payment, and electronic coupons as a means of marketing.

To link the valuation of countless coupons in the form of OIC, an abstract coupon or point unit to serve as the basis of valuation is necessary: the hypothetical generic point (GP). GPs do not actually have to be issued as cash and could also be designed in the form of a product basket. If a certain GP were issued in relation to the total amount of OIC, this GP could be extremely powerful as a means and unit of exchange and settlement. All exchange rates with the currencies of each country and the exchange rates of respective OIC would be simplified through this GP. Replacing this with the currency of a designated nation state would not be consistent with our idea; an electronic transaction platform of electronic OIC would probably issue GP. In any case, for OIC to be used as a means of payment and settlement, an intermediary unit such as GP would play an important role.

A secondary bond market is a system with quite a high information-processing load, including management of the valuation of a diverse range of companies. Exchanges of macro-bonds are now rapidly being constructed in the form of electronic bond exchanges. Furthermore, as a result of the application of information technologies, bond markets are becoming more liquid. However, because the electronic platform for the OIC we envisage presupposes a payment and settlement system using OIC, it could not be replaced with the platforms of existing bond markets. For OIC to be possible, it is necessary for an exchange market platform to be established that allows pluralistic organization valuation and the exchange of OIC. In that case, a platform is essential that allows countless types of open loop electronic money that expresses OIC on information networks and that allows conversion between these freely. In addition, the market for the OIC we foresee must have a different spread from conventional markets for institutional investors. It is not certain whether such an electronic platform for OIC could kick off as a business. However, for companies and various organizations to be able to receive a credit supply based on the added value they themselves create, such as system is inevitable.

### 4.5.2 Scenarios of the Growth of OIC

Organization-issued and company-issued cash could only become a global means of expression, saving and exchange of value, and issue and circulation of credit on the assumption of electronic networks. However, OIC will for the time being no doubt start from business models where the aim is the exclusive winning of customers. There is the possibility that before long, as these link up and acquire transferability, depending on circumstances they will form business models that allow open exchange and form enormous networks for the expression, saving and exchange of value, and the creation, supply, and circulation of credit for a new era. More than anything, this system will be a system that evaluates the added value production activities of various organizations and individuals on the basis of various valuation landscapes, including not only market but also nonmarket network valuation, and then creates and supplies credit. This is a type of decentralized climax of systems of credit creation and supply and has the possibility of very fundamentally changing the nature of existing socioeconomic systems. However, for such a system to actually grow and be accepted, concrete, strategic scenarios are necessary.

#### *4.5.2.1 Scenario of Platform Growth*

The first scenario where a platform of OIC could grow is the possibility that leading companies create a consortium and embark on the issue of OIC. For example, if leading customer-oriented companies such as Sony Corporation or Microsoft Corporation issued OIC to price their own products this would have a great impact.

The second scenario is the possibility of companies that issue through conventional point systems creating a secondary market for these points and coupons. In such a case, because an interchange market of diverse coupons and points including high street coupons would be formed, many issuing bodies of coupons and points would take part. However, in reality there are disadvantages and realization would not be easy; individual point and coupon systems are separate and not based on transferability and that for each company, because a transfer market is created, the effectiveness of incentives to keep customers for their own sales platform is reduced.

The third scenario is game company-issued money, useable in games and linked to actual money. This is something that could easily occur when a next-generation network game can gain one million users and becomes a powerful platform for virtual customer malls. In the world of network games, we are familiar with such virtual currency and there is the possibility of OIC being issued in this world of games for pleasure (see http://www.gamingopenmarket.com/). Another similar possibility is for the ball-saving system in Japanese pinball or pachinko to become generalized and OIC. For example, the ball-saving system in Daikoko Denki's pachinko is a type of electronic money and is partly in accord with the outline of this business model of OIC (see http://www.daikoku.co.jp/). Depending on your viewpoint, Daikoku Denki Corporation's ball-saving system (RoboCard) is the largest scale card-type electronic money currently in existence (see http://www.daikoku.co.jp/index.html).

The fifth scenario involves bond electronic market platform operators, banks, credit card companies, and other players operating OIC as a part of electronic financial products. The sixth scenario is the possibility of market creation by companies that provide servers of OIC; where providers of OIC exchange market servers conduct business with the issuing systems of OIC and its exchange market servers. In the meantime, expansion of OIC and the demand for exchange market servers that perform relationship building and exchange will no doubt greatly benefit companies that take the initiative over electronic systems for OIC. The seventh scenario is for manufacturing industries such as manufacturing industrial concentrations to form consortiums, express in OIC their own processing ability as decentralized manufacturing servers, and issue OIC as a means of saving and exchanging that value. This links with the possibility of creating a new industrial structure by linking with the scenario of on-line manufacturing that makes open-source manufacturing designs (Deguchi 2002). The eighth scenario is the possibility that under government leadership a trial of this new finance system will commence. One possibility is an initiative to conduct an OIC trial using an extension of government-supported community cash like ecomoney.

It is possible that these scenarios will proceed, not independently, but in a form where several scenarios blend. In the long term, in a world with a high level of information processing ability based on electronic networks, one climax of the nature of money will be OIC spreading as a form of cash that mixes incentives to customers, credit creation (direct finance), and means of payment settlement. The following scenarios for materializing OIC consider the steps by which such a system of OIC would grow:

1. Mixture of systems with a partial cash function.

   As community money and coupons, point cards, and prepaid cards become electronic and form networks, their boundaries will gradually become unclear. At this stage, such common characteristics such as issuing systems will increase and services will emerge where multiple systems are mixed and shared.

2. Development of a shift to open loop.

   In these processes, coupons and point systems that were previously nontransferable and closed loop-type will switch to open loop, allowing them to circulate repeatedly on networks. The basic technology behind electronic OIC that has been made open loop is virtually the same as the technology of open-loop electronic money. This means that the technology of electronic money is also available to OIC-issuing businesses allowing many issuing servers of electronic OIC, including electronic coupons to start up. Exchange market servers between these OIC then become essential.

3. Growth of OIC-issuing servers and the exchange market server business.

   With electronic OIC, it is essential to enable operations such as the interchange and transfer of OIC issued by many organizations, and to allow this, some type of open standard is required. Another requirement is for a variety of additional services to evolve. If server software is available free of charge for the small-scale issue of OIC, it will be extremely easy for small companies and organizations to have a local prepaid system or electronic OIC, such as coupons or gift tokens. A business model where the issue of OIC up to a certain scale is free of charge is essential for the successful spread of such systems. It is important to include related areas, such as value-added services relating to the mainstays of profit, server agency business, and security services; in particular, exchange market servers for the intermanagement of diverse OIC will no doubt become a major business and serve as a profit center.

4. Donations using electronic OIC.

   As a result of an increase in the use of coupon-type OIC for donations there has been a progressive development of the coupons exchange market and a consequent expansion of exchange business. Support from a taxation perspective such as a donation tax system, where taxpayers directly select the recipient organization and use OIC to make their donation, are the test case that will predict the future possibility of decentralization of credit creation and supply.

5. Development of exchange markets between electronic coupon systems and regular money.

   When a vast secondary market for OIC has been created, its exchange service will generate adequate commission income and in addition, management services of OIC with a variety of added value will also be feasible businesses. An exchange market between OIC will be a kind of virtual note shop and at the same time an exchange market of virtual currencies. Through the start up of exchange markets between virtual currencies and actual legal tender, electronic coupons and electronic OIC will come to have the essential function of cash.

6. Development of diverse financial product design for speculation and investment and credit creation and supply.

   A few years into the future and the OIC in circulation will have a redemption mechanism based on a variety of financial product designs, for example, company coupons to purchase a company's products and services and investment coupons for movies to raise the production costs. If the coupons of a new movie or musical production were purchased at low cost, their value may rise before the opening to the price of several seat tickets; a prime example of speculation. In this way local decentralized investment is possible through electronic OIC. On the other hand, various nonprofit organizations and their activities and services could also be evaluated using OIC and put into circulation in various ways besides their redemption as a donation.

7. Variety of financial product designs for OIC.

   Let us assume that a company product or service worth 100 yen has been priced at 90 points of company OIC. If at that time on the market 90 points can be purchased for 98 yen, consumers will no doubt buy the product or service with points. The company itself can also issue coupons where this difference between the face value price and selling price is an incentive. Alternatively if an intermediary business such as a note shop sells coupons to consumers for the company, this means that the shop has the function of a selling agent in place of the company. Besides being issued in the form of discount bonds, it is also possible for OIC to be issued in the form of so-called coupon bonds. In this case, OIC becomes a type of points bond where after a certain date, point interest is added to the initial value. Even with OIC that allows immediate redemption for services or products, as seen in the example of prepaid cards, there is a time delay between the issue and the redemption of products and services. There are benefits to hoarding and an average redemption time creates the same effect as holding bonds. For example, in the case of prepaid telephone cards, if the transfer of points and transfer with other points could be carried out instantaneously on a network and if telephone cards can be used for the payment of telephone charges, this would allow the possibility to pay telephone charges on the network. OIC needs to be designed in the form of financial products that are flexible enough to allow such diverse possibilities.

8. Development into a more stable system than nationally issued cash.

   With the development of an exchange market between OIC, the local credit issue mechanism using OIC will gradually grow into a vast system. As a result, even if a crisis of confidence occurs locally due to corporate bankruptcy for example, over a wide area a system will be formed that grows steadily and is more stable than networks of nationally issued legal tender.

## 4.6 Pluralistic Valuation and the Construction of a Cross-Referential Society

The formation of social networks with various values that develop together on the Internet means the emergence of a new electronic public sphere (Deguchi 1999). Ideally, when a social network evolves around an issue and a network valuation landscape is formed, more than one such network valuation basis (landscape) will be formed and by seeking support these will compete and develop together. It is possible that the valuation and competition of valuation landscapes will give rise to a dynamism that transcends capitalism in a game of competing network valuations.

This is why organizations and a variety of mechanisms for valuation, reference, and guarantee must grow as the platform on which this new network public sphere rests; organizations for the valuation and rating of NPOs, etc. in various forms are already starting to develop (Tanimoto 2000).

To make the network public zone sound, the formation of a cross-referential society where the various activities of organizations are mutually evaluated and compared is inevitable. To enable competition between valuation landscapes, it is necessary to introduce dynamism whereby the bodies that evaluate and are involved in resource distribution—social networks, approval, valuation, rating, and guarantee institutions—are also subject to network valuation and must also compete. This is because if valuation, as in the past, is limited to a small number of financial institutions, public approval, and government, the dynamism of the system will be lost.

Because more than one person makes an evaluation, markets selling products and services have a structure where pluralistic valuation bases are layered on top of each other. As a result, the wishes of the people who use special products and services reach the market through the size of their demand. However, on stock markets and bond markets that conduct credit supply for the activities on profit-making companies, even if the valuation by individual investors is different, the valuation landscape of profit maximization is the same. Stock and bond markets cannot adequately perform the function of a social circuit that links the wishes of people who would like to use special products and services with a small market scale and profit, to the supply of credit to the organizations that produce them. Stock and bond markets are designed as a game aimed at profit or capital gain maximization. Of course, provided there is large enough demand, the game of profit maximization links with demand based on people's pluralistic values, but this link is not always strong. Even if on markets selling goods and services, there was pluralism in the valuation of products, corporate valuation on capital markets with the role of supplying credit for the capital investment of the companies that make them is still carried out based on profit maximization or capital gain maximization. Between these two markets there is a loose link, but this link is not close and furthermore resource distribution via the market does not function normally in relation to a type of public service like volunteer activities or the tackling of environmental problems.

In this respect, it has previously been the understanding that it is the state's role to make up for the failures of the market and secure the activities of nonmarket organizations. However, as already noted, uniform valuation and resource redistribution by the state has several problems. To begin with, the valuation landscape changes slowly and valuation is prone to be static or distorted. Moreover, on the assumption of small government, it is difficult to bring together within bureaucratic organizations the experts who could properly evaluate various modern-day issues. In other words, it needs to be understood that even if the government is the body that conducts system design for valuation and resource distribution, its function of carrying out resource distribution based on the evaluation of individual areas is limited. Even if the organization of the state were enlarged, this would not help in the formation of pluralistic valuation bases (Deguchi 2004).

To introduce dynamism of valuation and choice and resource distribution different from the so-called market solution, OIC could be used. For example, in relation to NPO activities that provide common services or shared goods evaluated using some kind of value basis, a system could be designed that allowed people to purchase OIC issued by the NPO. The system could allow each person to choose whether, based on an ex post facto evaluation after the activities, they actually redeem this as a donation or they request redemption of those funds in some other form. To allow such a transaction, a process like a donation tax system is required. Taxpayers who have paid donation tax in the form of purchasing the OIC of an NPO engaged in activities that provide a service they themselves consider important, would be able to choose to redeem the OIC as a donation, either by making their own valuation of the activities after a certain period, or entrusting this to an evaluation organization. If a choice were made not to make a donation, the NPO in question would be in the situation where it has a liability to the government and would be forced to redeem this in some other form. Such system design concerns an area where economics with a conventional market-oriented global outlook has not performed in the past. However, we believe that at this turning point it is an important role of economics to design a system for this new age and present scenarios for its realization (Deguchi 2004).

At a time when valuation landscapes themselves are in competition, it is necessary to reconstruct a mechanism for regulating and resolving the conflicts that arise between the activities of various organizations. On the one hand this will take the form of competition to win resources through evaluation of organization activities and evaluation of the valuation landscapes themselves. However, on the other hand there are also expectations of the role of government in such a process. The government will no doubt need to gradually evolve into a flexible system that supports adjustment and coexistence between various value landscapes and regulates conflicts.

An age of ultrapluralistic money based on pluralistic values is possible for all kinds of organizations that form added value and issue OIC, but it is unthinkable without IT. It may not then be the task of economics alone, which gives explanations only on the basis of the system of the market, to talk about a system of money that does not yet exist. However, on the other hand, if we consider the history of money objectively, it is by no means unrealistic that myriad OIC networks are one of the ultimate shapes the economic game will have in the future. Our future depends on what we envisage and the kind of system we require.

## References

Deguchi H (1994) Networks (in Japanese). JUSE, Tokyo

Deguchi H (1999) Global management of international public goods (in Japanese). Annals of Public and Co-operative Economy 9/10:110–119

Deguchi H (2002) Open source manufacturing on industrial agglomeration (in Japanese). Soshiki Kagaku 23:22–28

Deguchi H (2004) Economics as an agent-based complex system. Springer, Berlin Heidelberg New York

Iwai K (1998) Money theory (in Japanese). Chikuma Shobo, Tokyo

Kato T (2001) The new century of ecomoney — evolving twenty-first century economics and society (in Japanese). Keiso Shobo, Tokyo

Solomon LD (1996) Rethinking our centralized monetary system: the case for a system of local currencies. Praeger, New York

Menger C (1984) The origin of money. Committee for Monetary Research and Education, New York

Tanimoto H (2000) Transcending the boundaries of NPOs and companies — the commercialization of NPOs and NPO-type companies (in Japanese). Soshiki Kagaku 33:19-41

# 5 Strategic Alliances and Innovative Performance in Network Industries: The Case of the Mobile Internet in Japan and Europe

Michael Haas[i] and Franz Waldenberger[ii]

i Japan Centre of Munich University, Oettingenstr. 67, 80538 Munich, German, mhaas@bwl.uni-muenchen.de
ii Japan Centre of Munich University, Oettingenstr. 67, 80538 Munich, German, waldenberger@bwl.uni-muenchen.de

**Summary**

We argue that the creation of strategic alliances led by mobile network operators has been decisive for the promotion of mobile Internet services in Japan. Exclusive business alliances allowed for the internalization of externalities resulting from the systemic nature of the innovation process. This led not only to an improved technical quality of mobile Internet systems, but also to a more efficient organization of content supply. In the early stages of market development, these benefits contributed to a fast development of demand and more than compensated the possible disadvantage that exclusive business advantages cannot fully exploit economies of scale.

**Key Words** Mobile Internet, Network industries, Strategic alliances, Relational Dominance, Exclusive Business Relations

## 5.1 Motivation

How can we explain the success of mobile Internet communication in Japan in view of its failure in Europe? Generally speaking, there are many possible reasons, for example:

- Differences in user preferences

- Differences in the diffusion and costs of the fixed Internet
- Different approaches in the management of the innovation process

In this chapter, we focus on the third of these possible explanations. The introduction of the mobile Internet was confronted with four sets of problems. First, it was a systemic innovation requiring the coordinated research and marketing efforts of network operators, handset manufacturers, and providers of on-line content. Second, there was fundamental uncertainty about the needs of potential users. What content would be able to best exploit latent demand? Third, in organizing the supply of on-line content, the peculiar cost structure of media products had to be taken into account. The production of content is characterized by relatively high fixed cost for the "first copy," and relatively low, almost negligible variable costs resulting from the consequent use of content. Such a cost structure precludes the organization of content supply in competitive markets where price is set equal to marginal cost. Fourth, and related to the cost structure, attention had to be paid to the positive feedback loop between content and demand. More content variety increases the value of the system and therefore the user rate, which in turn lowers the average cost of supplying content.

The Japanese approach to managing the innovation process was fundamentally different to the European approach. In Europe, the main players in the industry pursued individual strategies. This resulted in free competition in all parts of the system, i.e., at the level of network operators, handset manufacturers, and providers of on-line content. In the Japanese case, each network operator formed exclusive business relationships with a selected group of handset manufacturers and handset browser companies. Within their strategic alliance, network operators actively structured and organized the market for on-line content. In our view, the Japanese approach was not only able to achieve a better technical quality, it was also better suited to resolve demand uncertainty, to account for the inherent failure of competition in the on-line content market, and, as a consequence, to make use of the positive feedback loop between content variety and demand.

In Sect. 5.2, we explain why unstructured competition among software and hardware producers can fail in developing an innovative media market, like the mobile Internet, and how such a failure can be avoided by strategic alliances among suppliers of system components. In Sect. 5.3, we show that Japanese network operators in fact followed a strategic alliance strategy to develop the market for mobile Internet. In Sect. 5.4, we take up the question of why Japan and Europe chose such different approaches. Section 5.5 discusses the further evolution of the mobile Internet in Japan and Europe, showing that European players significantly modified their strategies partly imitating Japanese approaches. In concluding, we discuss theoretical and managerial implications of our observations.

## 5.2 How the Restriction of Competition Between Component Suppliers Could Speed Up the Transition Toward the Mobile Internet

The problem of developing the mobile Internet can best be approached by looking at the various factors that determine the demand of users. The rate of use per period will depend on:

- The level of prices
- The technical quality of the system
- The variety of content
- The accumulated experience of past use

We focus on two of the four factors pointed out above. These are the challenges of achieving a high technical quality of the system and variety of content, respectively. We then discuss how the restriction of competition between component suppliers contributes to overcome these problems.

### 5.2.1 Coordinating Development of a New Technical System

To assure a satisfactory technical quality, network operators and handset producers must not only invest in their own technology, but their development efforts must also be coordinated with regard to technical specifications and timing to ensure the compatibility of networks and handsets. Last, but not least, quality and reliability of the system require that the hardware components of the system are able to access and run the software supplied in the form of on-line content. Given the novelty and the complexity of these tasks, comprehensive testing is needed to assure the functioning of the integrated system.

### 5.2.2 Organizing Content Supply

The development of a technical platform is only one step toward the mobile Internet. The on-line content defines the functional scope of the system. It comprises information, such as news, pictures, music, or videos, but also includes interactive entertainment, communication, and transaction functions. Although, technically speaking and in terms of costs, content might be considered to represent the least complex and least expensive part of the system, it is in the end the part that induces costumers to use and pay for the mobile Internet. Users care little about the sophistication of the technology; they are only interested in what the mobile Internet offers them in terms of functions and convenience, including design and user friendliness.

The provision of content variety is confronted with two problems. First, there is a fundamental uncertainty about whether new content will be accepted by users. This can only be determined after the content has been introduced. Although this is basically true for all innovative products, the problem is particularly severe for media products such as on-line content, because the prototype or "first copy" consumes almost all of the total cost and because it is in many cases not possible to make ex post adjustments to the product to better fit demand. This means that most of the costs for providing content have been "sunk" before demand is known.

Second, due to the specific cost structure of media products, free price competition will not work. If the price elasticity of demand is high, prices will be close to marginal cost, which means they will be below average cost and will preclude the financial viability of content supply.

Finally, in new markets, consumers have to learn about the innovative products. This is depicted by the fact that actual demand depends on accumulated past user experience. The demand interdependence will be strongest in the early stages of market development. This implies that the effects of quality and content variety will not only be felt in the level of actual demand, but will also influence the growth of demand. Small temporary mistakes can thus have a strong impact on the development of the market.

### 5.2.3 Benefits of Restricting Competition

The question to be discussed is: will unstructured competition among mobile network operators (MNO), handset manufacturers (HM), and independent content providers (ICP) or competition among strategic business alliances lead to a faster development of the mobile Internet market? Unstructured competition avoids the replication of investment that occurs in the context of competing business alliances where content and handsets have to be developed within each competing group. This form of competition is thus better able to exploit economies of scale. On the other hand, free competition performs badly in an environment in which the actions of economic agents are accompanied by strong externalities due to incomplete property rights and high transaction costs, and it does not work at all when it pushes market prices below average cost. In such an environment, the restriction of competition within exclusive business alliances can improve performance, because it allows at least for a partial internalization of externalities and it lowers the price elasticity of demand. The benefits bear relevance to both the technical quality of mobile Internet systems and to the organization of content supply.

For a situation in which every HM cooperates freely with any MNO, an HM and an MNO will have low incentives to invest in the improvement of the technical interface between mobile networks and handsets, because it will be difficult to secure the resulting benefits. In principle, the jointly developed technical improvement could be patented and licensed without restrictions on business relationships. However, the licensing of untested technical solutions is burdened with high transaction costs that in turn could only be avoided at the expense of knowledge spillovers. A cheaper solution of protecting the benefits of joint technical improvements can be achieved if both parties agree on the joint exclusive use of the improvement for a specified period of time. The implied exclusive business relationship prevents competitors from using the technical advances,[1] and secures a temporary market lead that might even perpetuate itself in a fast-growing market. Because the quality of the mobile Internet depends on the reliability and technical sophistication of the interface between mobile networks and handsets, and because there is room for improving and fine tuning of the interface in the early stages of the mobile Internet, exclusive business relations between one MNO and a selected group of MNs will lead to a better-quality and faster market development. The ability to better fine tune the interfaces is further enhanced by the fact that the cost of testing the compatibility of components is reduced because the number of browser–handset–network combinations to be tested is limited by the size and composition of the respective alliance.

It has already been argued that the provision of on-line content cannot be provided in a competitive market environment. To cope with the problem of fundamental demand uncertainty and to avoid marginal cost pricing, media markets tend to bundle content. Examples of content bundling are newspapers, music labels, and book publishers. Bundling of content cannot overcome fundamental demand uncertainty, but it can pool the associated risks. Bundling also allows for joint marketing activities, investment in brand reputation, and for differentiation along the variety dimension, thus lowering the price elasticity of

[1] Such exclusive business relationships are common practice in the joint development of parts between, for example, car manufacturers and parts suppliers.

demand. Last, but not least, the management of bundled content is better able to control quality and to internalize resulting benefits.

In principle, the various advantages of bundling content can be exploited by independent providers of mobile Internet services and within large media companies. Why should ICP be part of the strategic alliance of MNO and HM? Our basic answer is that: (1) the integration of content into a portal site owned by the MNO can benefit from joint uses of "resources," and (2) in the early phases of market development such a joint use of resources is difficult to arrange through short-term contractual agreements. The resources to be jointly used by the MNO and ICP are the reputation of the respective MNO, its communication channels and billing relations with existing subscribers, the joint efforts of the MNO and HM in developing and specifying handsets and portals, and the marketing of mobile Internet services. Because the use of content implies the use of a mobile network and a handset, marketing of services will increase the use of mobile networks and the demand for handsets. From the perspective of users, one portal site and one billing partner reduces transaction costs and enhances the convenience of the system.

The advantages in favor of exclusive business alliances have to be weighted against the disadvantage of foregoing economies of scale. In the early stages of market development, the advantages will generally outweigh the disadvantages, because technical quality and functional variety are essential to get the market started in the first place.

## 5.3 Developing the Mobile Internet in Japan and Europe

This section provides qualitative evidence on how Japanese and European players of the mobile telecommunications industry expanded the system architecture of so-called second-generation mobile telecommunication systems during the late 1990s in order to facilitate mobile Internet services.

These systems had been developed through the 1980s and had replaced the first generation of analog mobile telecommunications (in operation since the late 1970s) at the beginning of the 1990s. Several standards specifying the architecture and the interfaces of these systems had been issued and internationally diffused. Among the alternatives, the Global System for Mobile Telecommunications (GSM) emerged as the most successful standard. Most European countries, as well as other nations, had followed the decision of the 16 EU member states to adopt GSM as a national standard. A total of 167 network operators within 103 countries ran GSM systems commercially by 1996 (Funk and Methe 2001). In contrast, Japanese network operators had chosen to deploy systems based on the Japanese Personal Digital Cellular (PDC) standard as well as the Code Division Multiple Access (CDMA) standard developed by the American company Qualcomm. While CDMA was relatively successful concerning its international diffusion, no country other than Japan adopted PDC.

While upgrading second-generation systems, technical suppliers and network operators proceeded with the development of third-generation systems that were thought to provide a superior infrastructure for mobile Internet services and to ultimately replace the old systems. Three standards for these new systems were issued in the year 2000: an evolution of GSM called W-CDMA, ANSI-136 (dubbed as UWC-1367), and ANSI-95 the antecessor of CDMA called cdma2000 (Kano 2000).

The following subsections compare, however, only Japanese and European approaches toward upgrading the second-generation systems. Approaches in innovation management with respect to developing third-generation systems are outside the scope of this chapter. The data presented in the following sections were gathered in the course of a comparative case study. The study followed the methodology as proposed by authors such as Yin (2003) and Eisenhardt (1989). A total of 27 managers and leading engineers were interviewed between 2002 and 2004. The interviewees were employees of network operators, handset manufacturers, software developers, and content providers. All of the interviewees had been directly—or at least closely—related to the development activities. Each interview lasted between 1 and 1.5 h.

### 5.3.1 Establishing and Leading Alliances for the Development of Technical Components: The Japanese Approach

At the end of the 1990s, five groups of MNO were competing in the Japanese market for mobile telecommunication services. NTT DoCoMo, the largest among the MNO, established the first mobile Internet platform –called i-mode– in February 1999 (see Funk 2001; Natsuno 2000). DDI and IDO (both later merged into KDDI establishing the brand name au) as well as J-Phone (now Vodafone) followed suit in April and December 1999. Although smaller in size, they were also able to set up profitable businesses for mobile Internet services. All three Japanese MNO differ not only in size, but also in strategies they pursue and in the set of services they offer. However, in one aspect they are alike: they all led and dominated a strategic alliance to develop the market for mobile Internet services.

First indications of the dominance of Japanese MNO can be perceived in the development of technical platforms (i-mode, Ezweb, J-Sky). The MNO initiated and strictly controlled the innovation processes necessary to develop handsets and browser software, which are the two most important components of the system hardware. In the early stages of the development project, only the MNO, one or two pilot handset vendors, and (in two out of three cases) external software developers were involved. MNO, pilot vendor, and software developer jointly developed the necessary technical specifications of the particular mobile Internet platform. The other HM, being a partner of the particular MNO, later adopted the newly developed and tested specifications, thereby enlarging the alliance. In all cases, MNO had the final decision rights concerning what specifications HM should deploy. Testing of the different components and enforcement of specifications were also carried out by the MNO.

In mid-1997, DoCoMo set up two project teams to test the technical feasibility of mobile Internet browser software and to work out the necessary interfaces.

Each project team was composed of one HM (NEC and Matsushita respectively), ACCESS (a supplier of browser software for constrained devices), and DoCoMo itself (Shirakura 2002a). While DoCoMo specified in detail how the user experience should look and controlled the requirements to be fulfilled, it did not prescribe how to achieve the set goals. After successful completion of the first step, DoCoMo convinced two more HM (Mitsubishi and Fujitsu) to join the development process and established two more project teams. It is noteworthy that DoCoMo restricted information flows between the teams (Shirakura 2002b). Keeping the teams uncertain about the progress of the others created, on the one hand, a competitive environment, which stimulated development efforts and gave handset producers enough freedom to come up with different solutions. On the other hand, as DoCoMo was present in each team, it was able to intervene and coordinate the different technical solutions, thus preventing the emergence of interoperability problems. Three out of the four HM, for example, had decided to license the browser software developed by ACCESS. Only Matsushita opted to develop its own browser solution. DoCoMo, however, demanded that the look and feel of content should be the same on any handset, so that Matsushita had to struggle to achieve an output that was almost identical to that of the ACCESS browser. In the end, handsets were supplied from different handset manufacturers, but were nevertheless fully compatible with each other, offering the same look and feel of the i-mode service. The technical features of handsets relevant to content providers were so highly standardized that the different types could be thought of as being identical. This device standardization lowered content development costs, because only one device had to be considered and there was no need to create and test content on every handset type available (Pavlovski 2002). The exclusiveness of the strategic alliance is, according to employees of DoCoMo, limited to a certain period of time. After that period, technical specifications are made public. The period is, however, long enough to provide the members of the alliance with a head start that cannot be caught up by outsiders. Exclusivity of the alliance is continuously maintained as new development projects for updates of the i-mode platform are started in rapid succession.

Following DoCoMo, J-Phone, DDI, and IDO developed their own mobile Internet systems (J-Sky and EZweb) by following alliance strategies and achieved a degree of handset standardization comparable with that of DoCoMo. J-Phone developed the specifications in-house and cooperated with Sharp to improve and implement the specifications. Like DoCoMo, J-Phone did not specify any particular browser software, but enforced a similar look and feel of content.

DDI took the lead in developing the EZweb platform that was also adopted by IDO and the Tsuka Cellular Group. The approach of DDI differed from J-Phone and DoCoMo insofar as a foreign turnkey solution was implemented. The American company Unwired Planet (now Openwave) had developed a complete mobile Internet platform including network components and browser software. Unwired Planet was trying to sell its platform as a complete package and had managed to win AT&T Wireless, a US MNO, as its first customer. DDI decided to license the Unwired Planet platform in mid-1997 and required all its partnering HM to implement the Unwired Planet software. Hitachi and Kyocera volunteered as pilot handset vendors, but only the Hitachi handset was considered sufficiently mature to be launched together with EZweb services. Kyocera later introduced its EZweb handset together with other handset partners of DDI such as Sanyo who also had to license the browser software from Unwired Planet. Although the system was not developed from scratch, implementation of the Unwired Planet platform afforded considerable development work, as it had to be transformed from a Western language environment to one of Japanese.

A dominant position of Japanese MNO could also be observed in the branding of handsets. The brand logo of the MNO on mobile phones was predominant, while the brand of the manufacturer was not visible at all or only visible as a co-brand. It is thus fair to say that Japanese MNO sold the handsets rather as original equipment manufacturer (OEM) than as pure retailer to their subscribers.[2]

[2] A OEM is commonly defined as a company that uses product components from other companies to build a product that it sells under its own company name and brand.

The Japanese approach of developing technical components within exclusive alliances had certain advantages and disadvantages. The constant information exchange between the alliance partners allowed joint learning and quick readjustment of technical components leading to a fine-tuned system. By creating a unified system of browser software and handsets, Japanese MNO reduced not only development costs for ICP, they also enhanced the acceptance among users. Having reduced the number of different technical specifications, user learning costs and the probability of frustrating malfunctions (due to interoperability problems between different handsets brands) were reduced. A major drawback of the Japanese approach was that the technical components for each system were well aligned within, but not compatible across, system boundaries.

### 5.3.2 Creation of an Open and Standardized Platform in Europe

A strikingly different innovation process could be perceived in Europe. Unwired Planet was not only looking for customers in Japan, but also approached the then world-leading handset suppliers Nokia and Ericsson. Ericsson and Nokia, however, had already tackled the issue of developing mobile Internet platforms and had come up with their own innovative components. Thus, by early 1997, three competing technical solutions had emerged out of these uncoordinated innovation processes. According to Sigrudson (2001), Ericsson realized the danger of a "standard war" and took first steps toward the unification of the competing approaches.

By mid-1997 Ericsson, Unwired Planet, Motorola, and Nokia finally agreed on combining their different solutions into an open and standardized mobile Internet platform (later to be known as the Wireless Application Protocol or WAP). According to these four companies, WAP should not only be designed as an open and standardized platform, it was also envisioned to be deployable within any existing mobile telecommunications system no matter whether the particular system was based on GSM, CDMA, or any other standard. This initiative led to the establishment of the WAP Forum. The forum, however, remained closed to new partners until the first version of WAP (1.0) was completed. While the four founders still worked exclusively, other HM such as Siemens, Alcatel, and Samsung already announced support for WAP and later became new members of the opened forum. Even major information technology (IT) companies such as IBM and Microsoft joined the camp of WAP supporters. Established as an institution open to all parties, membership of the WAP Forum grew and reached 100 members in early 1999. Network providers also joined the WAP Forum but did not play an active role in conceptualizing and developing WAP (Sigrudson 2001). In early 1999, trials of WAP platforms were conducted all over the globe and the mobile telecommunications industry was very optimistic that mobile Internet services would gain mass acceptance. The trials, however, revealed serious flaws causing the members of the WAP Forum to issue an improved version of WAP (1.1) despite being close to entering the manufacturing stage (Hibberd 2000).

Although WAP had all the favorable prospects of a worldwide standard, the platforms finally built upon it were not as smoothly coordinated as in Japan. Members of the WAP Forum, especially Unwired Planet and Nokia, failed to agree on unified browser functionalities. Nokia and Unwired Planet both had developed their own browser software and were unwilling to compromise. Thus, browser functionalities relevant for the display of content were left optional. This led to a situation that is best captured by the following quote:

> One of the biggest problems WAP developers face is that their applications will be likely rendered in many different ways on different mobile phones. [...] Different browsers treat certain tags and commands in different ways, and myriad devices with the same browser also function unpredictably. Generally, this lack of coordination among handset providers, browsers, emulators, and the WAP standard creates a frustrating environment for even the best developers. (Kumar et al. 2003)

Consequently, content quality across all available handset types was hard to maintain. Valuable resources had to be devoted for testing instead of improving user experience. Content quality and usability were generally perceived as poor. Usability trials, for example, revealed that the users' improvements in the handling of WAP-enabled phones after 1 week of usage were appallingly low. This suggested that WAP was difficult to learn and was not very user friendly. Furthermore, some handset types, despite being sold as WAP-enabled, completely failed to display WAP content. Data transmission performance of WAP services was also perceived by customers as poor, leading to the sarcastic interpretation of the three-letter acronym: Wait And Pay.

Interview partners (both from HM and MNO camps) pointed out that lack of cooperation between the players also occurred with respect to testing and fine tuning of the systems. This phenomenon can be exemplified by the case of the German MNO Mannesmann D2 (now Vodafone) and T-Mobile. D2 as well as T-Mobile commissioned Siemens to integrate new WAP network components such as a WAP gateway into its operating GSM system. Unwired Planet in turn supplied the relevant components. Siemens had previously forged an alliance with Unwired Planet and also licensed the browser software for its handsets. While Siemens, Unwired Planet, and the MNO closely cooperated during the integration and trial phase of the new system, other HM stayed out of the proceedings. The platforms implemented at D2 and T-Mobile were thus only fine tuned with components supplied by Siemens and Unwired Planet. Other HM at first even refrained from providing handsets for trials. The other HM being Siemens' competitors feared that the network engineers of Siemens would turn the Nokia handsets over to their colleagues from the handset department. The MNO finally convinced HM to provide handsets for trials, but the mere provision of handsets is not what one could term end-to-end support. End-to-end fine tuning of handsets supplied by other HM would have implied that the handsets were accompanied by engineers helping to identify failures and maybe working on improvements at the site of the MNO. Without the engineers, the handset remained as black boxes to the MNO that could only report test results and hope that HM were listening to them. The same argument can be made for system trials that were made by Nokia. The tested systems were also exclusively composed of Nokia components.

Although Nokia and Unwired Planet had announced interoperability tests on a bilateral basis in February 1999, it has to be pointed out that these tests were limited to the interoperability of the gateways and the browser software supplied by both companies. Those bilateral tests, however, could not verify if, for example, a Nokia WAP phone and a gateway supplied by Unwired Planet still properly interacted when both were connected to a radio network produced by Ericsson. The same argument can be applied to other bilateral operability tests that were successively carried out in 1999. A centralized institution for end-to-end interoperability testing and for verifying that components complied with the WAP specifications was completely missing for about 1 year after the first WAP platforms entered commercial operations. Prior to establishment of the institution for interoperability testing and certification, HM could claim their products to be WAP compliant without providing any independent proof. The WAP Forum made a first step toward centralized interoperability testing in November 1999 by appointing the Open Forum Group as the certification authority for the browser functionality of handsets.[3]

Market introduction of WAP platforms was also badly coordinated with respect to timing of handset availability. For example, the acronym was often interpreted as "Where Are the Phones?" as handset suppliers failed to deliver upon the start of services by the MNO. Although content and networks were ready for service in summer 1999, users could not actually use WAP services, because handsets were only available in sufficient supply about 6 months later.

### 5.3.3 Making Independent Content Providers Alliance Partners: Organization of Content Supply in Japan

Japanese MNO did not limit their initiative to the development of an integrated platform; they were also proactively involved in the acquisition of content from ICP. DoCoMo went from door to door of potential content partners and convinceed 69 companies to supply mobile content at the launch of the i-mode platform. To motivate ICP to join into a cooperative arrangement, DoCoMo provided technical support for developing content and offered the bundling of

[3] http://www.wapforum.org/new/112299_open.htm

marketing and billing activities. Access to DoCoMo's customer base via a centralized portal site, use of the billing relationship and comarketing services were provided against a comparatively low handling charge, where DoCoMo kept 9% of content fees charged to the end users and transferred the rest to the ICP. Most content providers consider this arrangement as fair. Some analysts even claim that the handling charges are not sufficient to cover the costs of managing the portal site and billing relations, not to mention the comarketing services. However, the bundling activities are only available to ICP that underwent and passed a screening process by DoCoMo. Those official ICP are further inclined to comply with rules of conduct and quality standards specified by DoCoMo. Bad performance on the side of the ICP led to termination of the cooperative arrangement and loss of official status.

As indicated in Sect. 5.2, competition between official ICP was indeed controlled by Japanese MNO. Official status was only granted to applicants if they provided an innovative type of content not already available through other official ICP.

However, even official ICP did not benefit from airtime charges because these were fully taken by MNO. Risks and revenues were allocated according to the competencies an agent contributes to the platform. Official ICP only had to face the risk that development costs could not be recouped because the content in question failed to find its audience. Above all, this property rights structure provided ICP with strong incentives to keep up good performance in order to survive competition. Furthermore, the official vs. unofficial status of ICP posed a direct tool for quality control. DoCoMo clearly demonstrated that it did not compromise on quality by shutting out two of the initial content providers, thus launching i-mode with a content portfolio of 67 ICP (J@Pan Inc 2001). About 1 year after its launch, supply of i-mode sites had climbed from around 300 to 5500 sites. In the following year, the number of sites increased further by 600%, from 5500 to over 40 000 sites (Rangone et al. 2002).

It is noteworthy that DDI and IDO took a comparatively long time to provide billing services for ICP. In contrast to DoCoMo, which offered micro payment services right from the start of i-mode, au/KDDi introduced billing services several months after the start of the EZweb platform (at this time the merger between both companies had been completed). DDI as well as IDO had initially decided not to provide micro payment services for ICP. When DDI and IDO approached ICP asking for content compatible with their EZweb platform, ICP pointed out that DoCoMo had already contacted them offering better business conditions. Forced to match DoCoMo's offer, DDI and IDO modified their business model and started development of an adequate billing system. Because of the late start, the billing system could not be completed until the scheduled launch of the platform. To bridge the gap between platform launch and introduction of micro payment services, DDI and IDO compensated ICP by lump-sum payments (as a kind of development fee) in exchange for the content offered. Once the billing system was completed, au/KDDi, just as J-Phone had done before, fully adopted a business model similar to that of DoCoMo. The only difference between the companies is that J-Phone and au/KDDi keep 12% of content charges as the handling fee. In contrast to DoCoMo, both companies also guarantee to transfer content charges from subscribers that fail to pay their bills and therefore can charge a higher rate.

### 5.3.4 Keeping Independent Content Providers Out: Organization of Content Supply in Europe

While the Japanese MNO aimed at securing content supply by nurturing a semi-independent ICP community, European MNO opted at first for full integration of content provision. The MNO purchased the content from suppliers such as media companies and integrated the offering into their portal sites. Subscribers to WAP services in turn were only charged transmission fees.

ICP did not receive any support through comarketing or micro payment services. ICP willing to enter the mobile content market had to face the challenge of finding a way to earn revenues from their activities. In contrast to the fixed line Internet, revenues like those generated through advertisement were not feasible because the displays of the handsets were too small. Charging for content via credit card would have been possible, but it proved rather inconvenient to end users in comparison with billing via the telephone. Lacking a revenue source, it should not be surprising that the number of content suppliers rose only slowly (Rangone et al. 2002). It is, however, surprising that the content provided by MNO, which had full control over content production, lacked attractiveness. The initial content was, as one employee of a German MNO put it, of "creepy" quality. Computer magazines evaluating mobile content made equally unfavorable judgments (Kossel 2000).

## 5.4 Why Japanese MNO Could Create and Lead Strategic Alliances

In the previous sections, we provided qualitative evidence that Japanese MNO took a leading role during the development stage of mobile Internet services. Their European counterparts, however, did not play an active role in the conceptualizing and development of the WAP solution. The questions left open are, why were Japanese MNO able to establish and to lead strategic alliances, and how were they able to persuade HM to develop a unified MNO-specific design?

Answers to these questions are closely related to differences in the historical development of the mobile telecommunications industry in Japan and Europe. In its early days, the mobile telecommunications industry both in Japan and Europe had been dominated by publicly owned network operators that enjoyed a quasi-monopoly position and entertained strong and stable ties with domestic suppliers. In Europe, these business relationships, however, were completely replaced by arms-length transactions after the market for mobile telecommunications had been deregulated. In Japan, in contrast, business relations between MNO and suppliers remained exclusive despite the introduction of competition.

According to Funk (2003), the mobile division within the incumbent national operator NTT (spun out as NTT DoCoMo in 1992) could maintain exclusive relations with its suppliers because of its superior knowledge about the mobile communication standard PDC that it had originally developed (see also Kano 2000). HM such as Matsushita, NEC, Fujitsu, and Mitsubishi stayed loyal to DoCoMo, because the exclusive information provided by DoCoMo laid the foundation for a competitive advantage over other competing Japanese HM. In exchange for the preferential treatment by DoCoMo, the four HM tacitly agreed not to provide competing MNO with advanced handsets for a period of 6 months.

We do not contest this argument. However, we are not looking at market dominance, but are concerned with relational dominance in cooperative agreements between MNO, handset makers, browser developers, and ICP. Market dominance and preferential access to technological knowledge are both helpful when it comes to establishing relational dominance; however, we do not think that they constitute a necessary condition. For example, DoCoMo's competitors, lacking both market dominance and superior technological knowledge, were also able to establish relational dominance within their cooperative networks. Even with its superior position DoCoMo, could still have opted for a nonexclusive strategy.

The main challengers of the dominance of the MNO are HM. First, they are large enough to contest MNO dominance. Whereas most MNO still operate only in their home market, HM are competing globally. Second, they have a desire to differentiate their product from other HM, which means that they will resist demands by the MNO to submit to unified MNO-specific technical and design specifications. Third, as independent manufacturers, they will also be reluctant to become locked in with their product to one MNO. How could Japanese MNO overcome such resistance?

The ability of MNO to set up a strategic alliance rests upon the availability of various tools by which they can (1) commit to exclusivity and (2) induce HM to comply with MNO-specific technical and design standards. Let us first look at the possibility of establishing exclusivity. A rather easy but nevertheless important way to create exclusivity is by differentiating the standards of the interface between the mobile communication network and the handsets. By providing the relevant information exclusively to a selected group of HM, the MNO can grant preferential treatment. At the same time, the MNO makes a credible commitment to the exclusivity arrangement, because the possibility of switching to "outside" HM in the short run is withdrawn. By selecting more than one HM, the MNO preserves a superior bargaining position in the alliance.

Developing MNO-specific handsets implies that market demand is restricted to the demand of one MNO. This limits the number of products across which fixed development costs can be spread. In addition, HM are not only confronted with the general market risk that mobile Internet services altogether fail, they also face the risk that the MNO for whom their products have been exclusively developed and produced fails. To induce HM to enter into an exclusive business relation, the MNO must share in the development cost and in the additional risk. The MNO can do so in two ways: directly take part in joint research and development projects, or indirectly share in the development cost and in the risk by agreeing to purchase a minimum number of handsets at a prespecified price. Both strategies were indeed pursued by Japanese MNO (Natsuno 2000).

If relational dominance is a strategic choice, why did European MNO not adopt it? One straightforward and very simple explanation is the relatively large number of MNO compared with the small number of HM in Europe. By the time the development of mobile Internet solutions had begun, Ericsson, for example, supplied 280 operators worldwide. Developing MNO-specific components would have forced Ericsson to forego economies of scale. Ericsson had a very strong incentive to supply its customers on the basis of a single standard facilitating the most effective exploitation of scale economies (Sigurdson 2001). For a single MNO operating in a single country the necessary premium to convince Ericsson to enter a strategic alliance would have been prohibitively costly. Therefore, European MNO were not in a position to initiate a strategic alliance. Consequently, their involvement in the development of mobile Internet services was limited. In the Japanese case, two points are noteworthy: first, there were more HM than

MNO in Japan, and second, Japanese HM played only a marginal role in markets outside of Japan, suggesting that their performance was even more dependent on the supply to Japanese MNO.

## 5.5 Further Evolution of the Mobile Internet in Japan and Europe

Since the market launch of mobile Internet services, Japanese MNO have been engaged in a fierce battle to be the first to add new functionality to their platforms. In most cases, innovations were soon followed by competitors trying to top the new functionalities. The evolution of the mobile Internet in Japan continued in this fashion, finally leading to all kinds of electronic devices to be embedded into mobile phones. In Japan, mobile phones sold in 2004 encompass, just to provide a probably incomplete list, digital cameras, global positioning systems, analog TV, touchless smart cards, Flash clients, JAVA clients, e-mail clients, 40-chord polyphonic ring tones, high-resolution TFT color displays, external memory cards such as MiniSD, infrared and bluetooth interfaces, music players, AM/FM radio receivers, and fingerprint scanners. Apart from the continued expansion of second-generation systems, Japanese MNO also introduced third-generation systems. DoCoMo launched the world's first W-CDMA network in October 2001, while KDDi started commercial operation of CDMA2000 network in August 2002. The way Japanese MNO coordinated the underlying collective innovation processes did not differ from the approaches described in the previous section.

Approaches of European MNO, however, have changed dramatically. The last innovative component to be integrated into the European systems under the "traditional" management approach was multimedia messaging (MMS) functionality. MMS compromised the integration of digital cameras into mobile phones and the ability to send picture-enhanced and sound-enhanced text messages. After the spectacular failure of this innovative service caused by HM not being able to achieve interoperability of their components, European MNO increasingly moved toward the Japanese model, trying to establish relative dominance in their business relations with handset manufacturers (Massey 2004). At present, two strategies for achieving relative dominance can be perceived: (a) purchasing phones from no-name handset makers and marketing them under the brand name of the MNO and (b) differentiating the technical platforms and exerting bargaining power to convince handset makers to supply fully compatible handsets. Often both strategies are applied in combination.

In the past, European MNO relied mainly on the big global HM Nokia, Motorola, Samsung, Siemens, and Sony-Ericsson. These companies together cover 75% of the world market for handsets, each having gained a strong brand reputation with end users. To counter the strong marketing position of their main HM, European MNO have offered preferential distribution services to smaller Asian HM. These HM do not have a strong brand reputation and are therefore willing to sacrifice their own brand identity for preferential access to the MNO's sales channels. In addition, European operators have dramatically grown through international merger and acquisition activities. At the time of writing, Vodafone, for example, operates in 26 countries and enjoys a customer base of approximately 133 million subscribers. The balance of bargaining power, thus, has significantly shifted in favor of MNO.

A combination of both strategies can be seen in Vodafone's or T-Mobile's efforts to establish a new mobile Internet platform and in DoCoMo's attempt to introduce i-mode to the European markets. T-Mobile relaunched its earlier WAP platform under the new name t-zones in March 2002. In July 2003, Samsung was the first manufacturer to provide a JAVA-enabled handset exclusively tailored for new t-zones services. Since introduction of the Samsung phone and a no-name PDA supplied by a Taiwanese manufacturer, the t-zone portal encompassed also payable services. Content charges were due to an event-based billing scheme.[4] Vodafone used the know-how that it had bought by acquiring J-Phone to start a new mobile Internet platform under the brand name Vodafone live! Vodafone

[4] http://www.t-mobile.de/presse/1,1807,2562-_,00.html; cited 08.08.2004

promoted the new platform by providing three camera-equipped handsets that had been specially tailored to the new platform. Sharp, Panasonic, and Nokia supplied these handsets. O2 followed the strategy of its competitors and also introduced mobile phones (supplied by Siemens, Nokia, and Samsung) that were optimized for its O2 Active platform. O2 also made a bold move by introducing a monthly flat fee of 4.95 euro for unlimited data transmission.

Non-Japanese HM also changed their attitudes toward mutual cooperation even without increasing pressure from MNOs. Ways or institutions to facilitate cross-manufacturer interoperability testing have been established and the necessity to work more closely together in order to faster introduce innovative interfaces has been finally acknowledged. One factor driving this insight is the increasing competitive pressure from Asian component suppliers. These suppliers entertain cost advantages in the manufacturing process of standardized parts, but lack the technological know-how to develop innovative components. The only way for the "old" suppliers to survive against their low-cost competitors is to stay ahead by innovating and adding new features to the system (justifying higher prices). As the experience of WAP and MMS has shown, however, consumers are only willing to pay for new features if interoperability is granted.

The new approaches in innovation management and the change in the cooperative behavior of HM are slowly bearing fruit: industry experts (e.g. Scuka 2003), point out that European players have recently caught up in terms of technical sophistication of platform components. Drawbacks of the Japanese model are also becoming visible. It is common wisdom in the mobile industry —a point openly admitted even by Japanese interviewees— that the exclusive relations between handset suppliers and network operators lead to smaller batch sizes and thus higher handset prices. The European model resting on open standardized solutions allows, besides the cost advantage, for a broader participation in the innovation process. Although, for example, DoCoMo was the first network operator to introduce Java-applications, the quality of standardized Java-Applications has surpassed the quality of its proprietary DoJa (DoCoMo Java) solution, because the open model could attract a larger number of complementary developers. Put another way: the system architecture and the interfaces developed by European players finally seem to have sufficiently matured to unleash the benefits of a modular industry structure.

## 5.6 Conclusions

Exclusive alliances initiated and led by mobile network operators greatly supported the successful launch of mobile Internet services in the Japanese telecommunication industry. Strategic alliances under the leadership of mobile network operators provided handset manufacturers with strong incentives to invest in the fine tuning and optimization of respective technical interfaces. In contrast, the open character of the European cooperative arrangements failed to induce competitors to agree on crucial aspects of browser functionalities and gave only weak incentives to invest in a standardized mobile Internet platform beyond a basic level of commercial operability. European content developers, consequently, were forced to cope with the diversity in browser functionality and the overall lower quality of the technical platform. With fewer resources available for development of core content features, content variety remained limited and European developers were not able to achieve a content quality as high as their Japanese counterparts.

While the European industry has recently moved toward the Japanese approach of cooperative development, it should be born in mind that exclusive business alliances also have their drawbacks because they limit the market to the demand generated within the respective group, thus foregoing advantages of specialization. As standardization of interfaces proceeds and the industry matures, the advantages of specialization will increase and cost competition will intensify. We therefore expect a transition from strategic alliances to open competition during later phases of the industry's life cycle. However, the introduction of new generations of mobile communication networks might well interrupt these trends, as new standardization issues and demand uncertainties arise.

## Acknowledgments

We express our sincere gratitude to the interviewees for their collaboration. The Japan Society for the Promotion of Science (JSPS) greatly contributed to this study through the financial support of a 6-month research tenure at the University of Tokyo. We also thank Professor Osamu Sudoh and his team at the Interfaculty Initiative of Information Studies, University of Tokyo, for their cooperation and for providing a stimulating research environment.

## References

Eisenhardt KM (1989) Building theories from case study research. Academy of Management Review 14:532–550

Funk JL (2001) The mobile Internet. How Japan dialed up and the West disconnected. ISI Publication, Hamilton, Bermuda

Funk, JL (2003) Standards, dominant designs, and preferential acquisition of complementary assets through slight information advantages. Research Policy 32:1325–1341

Funk JL, Methe DT (2001) Market- and committee-based mechanisms in the creation and diffusion of global industry standards: the case of mobile communication. Research Policy 30:589–610

Hibberd M (2000) How it all began. Mobile Communications International December 2000 (electronic version), pp 1–3

J@Pan Inc (2001) Takeshi Natsuno. The Incrementalist. J@pan Inc 20, pp 34–38 u. 74–75

Kano S (2000) Technical innovations, standardization and regional comparison – a case study in mobile communications. Telecommunications Policy 24: 305–321

Kossel A (2000) WAP-Wahn, Das Märchen vom Surfen mit dem Handy. c't 9/2000, pp 194–195

Kumar V, Parimi S, Agrawal DP (2003) WAP: present and future. Pervasive Computing January–March 2003, pp 79–83

Massey J (2004) Critical mass or critical condition? Mobile Communications International May 2004 (electronic version), pp 1–5

Natsuno T (2000) Die i-mode Strategie: Warum der Rest der Welt nicht aufholen kann (in Japanese). Nikkei BP Kikaku, Tokyo

Pavlovski CJ (2002) Reference architecture for mobile Internet service platform. In: Proceedings of the 2nd Asian International Mobile Computing Conference (AMOC 2002). Langkawi, Malaysia, 14–17 May

Rangone A, Renga FM, Rafaello B (2002) Mobile Internet: an empirical study of the evolution of the supply of B2C mobile Internet applications in Italy. (available on-line at www.mobiforum.org/proceedings/papers/09/9.2.pdf)

Scuka D (2003) Mobile Kaizen and why Japan still matters.

Shirakura M (2002a) Before it was called i-mode - Part I (in Japanese). Nikkei Electronics 26 Aug, pp 203–207

Shirakura M (2002b) Before it was called i-mode - Part III (in Japanese). Nikkei Electronics 23 Sept, pp 243–247

Sigrudson J (2001) WAP OFF - Origin, Failure and Future. Working Paper, University of Edinburgh

Yin RK (2003) Applied case study research. Sage, Thousand Oaks

# 6 Value Creation on Networks and in Corporate Activities

Jiro Kokuryo[i] and Yoko Takeda[ii]

i Faculty of Environmental Information, Keio University, 5322 Endo Fujisawa, Kanagawa 252-8520, Japan, jkokuryo@sfc.keio.ac.jp
ii Yokohama National University, Graduate School of Environment and Information Sciences, 79-7 Tokiwadai Hodogaya-ku, Yokohama 240-8501, Japan, ytakeda@ynu.ac.jp

**Summary**
Information technologies enable the creation of new value in terms of changes in communication and cost structures. This chapter illustrates two cases in which a variety of knowledge of consumers or companies is pooled to produce new value: (1) a community site that has an enormous influence on product marketing, and (2) open-source software developed as a custom-made application system for a specific company.

**Key Words** Information technology, Value creation, Business model, Community site, Open-source software

## 6.1 Value Creation Through Information Technology

One of the major reasons for the effectiveness of information technology is that it facilitates changes to the fabric of business, society, and lifestyle. Information technology not only calculates symbols and transmits them, but also has a greater indirect effect whereby the manner in which it alters the flow of information facilitates the restructuring of organizational, industrial, and social systems.

Several of our existing systems are highly inefficient due to the restrictions on the flow of information. For example, as additional stages and levels are added to a growing organization, a chasm is formed between the top management and the ordinary employees. This then results in a situation where, although it would be more logical for the staff of overseas branches of companies involved in several businesses to receive instructions from the general manager for that business sector at the head office; in reality, branches have to operate under the instructions of branch managers alone, who may not have an in-depth understanding of the business. In scenarios such as this, the use of information technology like email or video-conferencing systems would again bring the top management and ordinary employees closer together and may enable the rebuilding of optimal chains of command for long-distance communication. This not only brings about direct results such as savings on transportation costs for meetings, but also indirect results that have a greater impact, such as better decision making through changes to the organizational structure.

The indirect effects of information technology can be roughly divided into two. The first effect is enhanced efficiency. In several economic systems, inefficiencies are generated by a maldistribution of information. In the present day, when price competitiveness on an international scale is crucial, there are extremely high prospects for efficiency enhancements through information technology. In several cases, clearing away information bottlenecks will enable the establishment of efficient operations under a totally different concept.

For example, in the truck transportation industry, trucks frequently return empty after delivering their cargo. Yet, in several cases, the slightest of detours would have ensured their carrying cargo during the return trip. However, the present information distribution cost structure has precluded the matching up of these trucks with available cargo. This situation results in inefficiencies because the next truck dispatched will also have to make the return trip empty. Hereafter, it may be possible to match up these sorts of needs in a flexible manner and at low cost by using a combination of the Internet and mobile phone-based email. Bringing the two together effectively will not only yield benefits to the truck and cargo owners, but will also generate savings for the society as a whole (Isagai and Kokuryo 2002).

The second effect is value creation. In order that the benefits of enhanced efficiency lead to revitalization of the economy as a whole, appealing products and services, on which consumers can spend the money saved through enhanced efficiency, need to be produced. Information technology contributes to the creation of added value by bringing together the creative strengths of different people and enhancing them. The key drivers will be the phenomenon in the industrial sector whereby modular products and technology are brought together to create things that did not previously exist, and also the phenomenon in companies where knowledge management leads to an emergence of knowledge. These phenomena will work to give rise to new industries.

In creating added value through information technology, the following two cases arise: (1) information technology is used for a product itself, and (2) information technology supports the mechanism for supplying a product. A classic example of case (1) are products that have an overwhelming appeal, such as mobile phones, and make consumers want to sacrifice part of their savings to buy them. There is a greater variety of possibilities for (2). For example, the manner in which word-of-mouth communication on community sites such as those dealt with in Sect. 6.2 have a great impact on product marketing. The systems wherein new products are being developed right from the design to the production stage based on data provided through information technology, such as open-source software, are dealt with in Sect. 6.3.

The current situation in Japan is such that while efforts are made to enhance efficiency, i.e., cost cutting, value creation does not progress at the same pace. By artfully achieving a balance between promoting efficiency enhancements and creating demand, we can bring about sustainable economic growth. Therefore, what Japan needs to do henceforth is quickly establish a mechanism for the creation of new value and to commit itself entirely to both enhancing efficiency and creating demand (Fig. 1).

**Fig. 1.** Enhanced efficiency and value creation

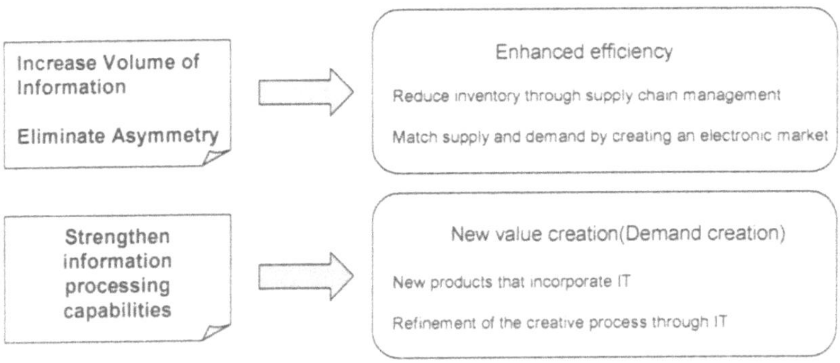


## 6.2 The Effect of Word-of-Mouth Communication on Community Sites

### 6.2.1 The Emergence of a Customer-Driven Business Model

The Internet has had a great impact on the cost of communication. There are two facets to this: (1) the absolute cost, and (2) the relative cost structure changes. Each of these depends on the technological aspects of the Internet (Kokuryo 1998).

In terms of absolute costs, this is typified by the dramatic fall in the cost of always-on connections in recent years. In 1999, low-speed 128-kbps circuits for always-on connections that serviced households and small business establishments used to cost more than 30 000 yen. However, in 2004, a 100-Mbps service is available for approximately 3000 yen. Speed has increased a thousand times, while the price has fallen to one tenth of its previous level. In terms of the cost per bit, this amounts to less than 1:10 000.

Another factor that is as important as the absolute cost is the changes in the cost structure. More specifically, it is the drastic fall in the cost of transmitting information from terminals compared with the cost of transmitting information from centralized sources. This is easily exemplified in the case of broadcasting technology, including TV. Radio wave-based broadcasting exhibits outstanding cost performance with centralized information being distributed to several people at a low cost. The system is characterized by the fact that even if the audience numbers increase, there is no increase in the fixed or variable costs; in fact, the higher the number of people watching, the lower the average cost. Under these conditions, in terms of mass broadcasting to the public at least, it is believed that the Internet will never be able to overtake broadcasting technology regardless of the extent of development. However, despite this fact, the cost of disseminating information horizontally at the user end remains extremely high. Broadcasting via TV and the like incurs huge fixed costs; therefore, it is only feasible for large organizations where large audience sizes are assumed.

A change in the cost structure for communication is highly significant because it induces structural changes in organizations. When the cost of transmitting information is high and the cost of receiving it is low, and you want to increase information processing efficiency; then the most effective solution is to gather information in one place that is peopled by staff having high levels of information processing capabilities to make decisions and to transmit information in a top-down fashion. In companies, this implies concentrating elite staff in the head office and issuing commands from there. Even if the ideal scenario is that all members of the organization share information and make joint decisions, it is easy to imagine why this was a difficult venture at a time when there were no electronic bulletin boards or mailing lists and when information had to be exchanged by telephone or conventional mail.

The structure of the market also changes. When it was inexpensive to transmit information from centralized sources and expensive to do so from individual entities, the practice of mass-producing products in large factories, marketing them in the mass media, and then selling them on a massive scale to an anonymous public was an effective one. However, in current times, the Internet and mobile phones have armed each and every customer with the ability to transmit information. The growing importance of customers disseminating information is accompanied by the growing power of both the infomediaries handling that information (Hagel and Armstrong 1997), and the interactive dialogue with these users (Levine et al. 2000). Thus, we see the emergence of a customer-driven business model.

### 6.2.2 The Roles of a Read Only Member (ROM)

In order to illustrate the phenomenon that exemplifies changes in the structure of communications, an empirical study was conducted with regard to the behavior of ROMs on networks (ROM: a person who does not post comments at interactive communication forums, such as on-line meetings, mailing lists, chat rooms, and so on, but merely reads the comments of others). The study involved surveying members of a cosmetics community website "@cosme" (Ogawa and Kokuryo 2002). In previous studies on consumer behavior in networks, these ROMs were generally considered a dead end; that is, they were the end point in the flow of information. However, upon further investigation, we observed that people who are merely reading do in fact convey that information to third parties (Fig. 2).

**Fig. 2.** Flow of information through electronic many-to-many media

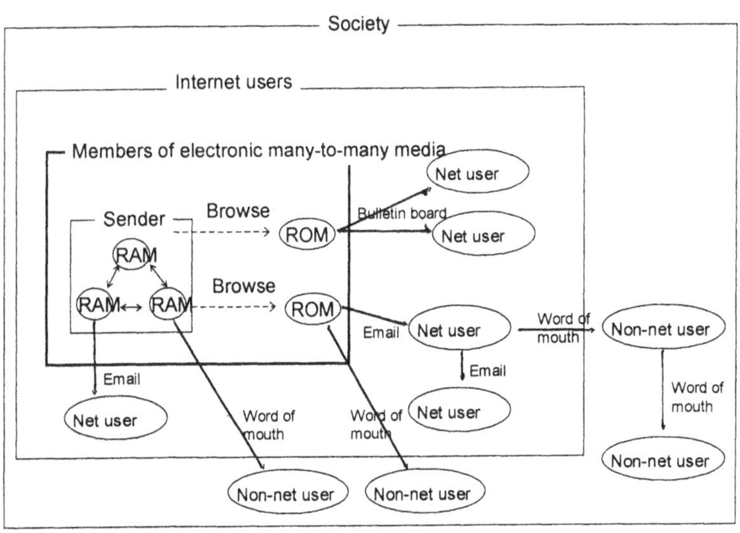

In the case of @cosme, more than 70% of ROMs communicated information posted by members to a third party (see Fig. 3), in addition to 93% of RAMs [RAM (random access member): a person who posts comments at interactive communication forums]. These figures are much higher than those obtained from other sites and are approximately double the figures garnered in studies targeting members of general sites. This difference is thus accounted for by the added value generated by @cosme.

**Fig. 3.** Percentage of @cosme members who shared information with a third party

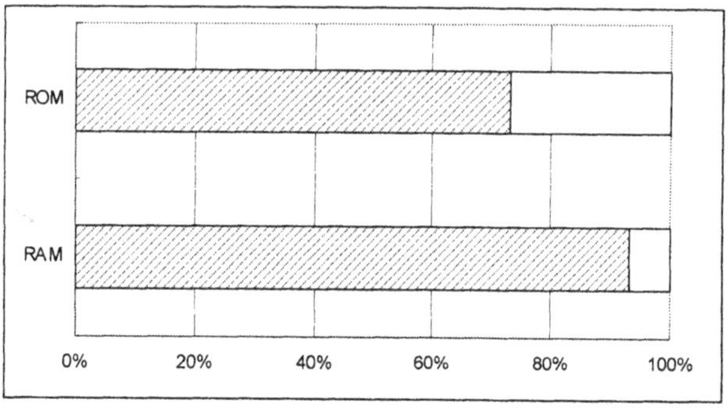


Another very interesting outcome of this study was that there was no significant statistical difference between ROMs and RAMs in their responses to the question, "Did the person you shared the information with actually buy the product, and is it fair to say that your sharing caused them to buy the product?" (Fig. 4) The fact that the ROM's influence is comparable to that of the RAM would suggest that the overall influence of the ROM layer as a whole is much greater than that of the RAM layer, simply because there are roughly nine times as many ROMs as RAMs. Even the most conservative estimates suggest that the ROM layer induces more than twice the number of purchases than the RAM layer (Ogawa et al. 2003). In other words, this implies that the influence of those people who are passive readers on networks is much greater than that of those who are active participants in total.

**Fig. 4.** Purchases by people who shared information on @cosme

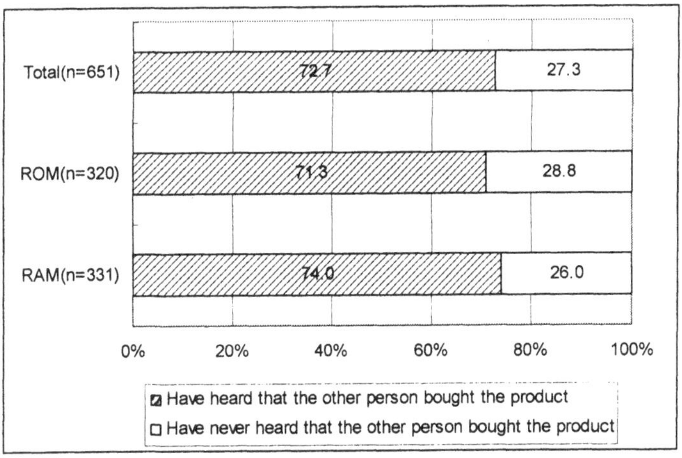


At @cosme, communication among members goes well beyond simple word-of-mouth communication, and there are cases where this has led to the development of new products. The lip gloss of one particular manufacturer was less expensive than the other brands, but it did not have a strong presence. However, after its reputation spread on @cosme as being inexpensive and of good quality, the sales increased dramatically. This company then conducted a popularity poll on @cosme, and based on those results it developed new colors. In doing so, it came up with a new hit product that has sold more than 2.5 million units in total (Ogawa and Kokuryo 2002).

The phenomenon in which an all-new value is created on networks has not yet been systematically demonstrated, and it will become an important topic for study in the future. Companies also need to organize how they should approach and respond to this network-based communication amongst customers. Although it has been recognized that the exchange of information amongst on-line customers has a very big impact on companies, the fact remains that this is not necessarily being incorporated in company activities. In several cases, the main reason for this is that the people who participate in bulletin boards and mailing lists still tend to have a "nerdy" image and are not viewed as potential customers, and that customers themselves refrain from commenting in forums where companies are involved, preferring instead to join in on customer-only sites (Kokuryo 1998).

## 6.3 When Companies "Open-Source" Their Own Systems

The development of open-source software, that is, software for which the source codes are made public so that it can be used or altered by anyone, is a classic example of value creation with various entities participating in intellectual production. Most open-source software has been developed for generic systems such as operating systems. Moreover, the developers are usually volunteer developers. However, in recent years, an interesting model has emerged in the restaurant and medical industries in Japan, where information system user companies and user organizations have paid information system developers to develop customized application software, and this software is later turned into open-source software (Takeda and Yoneyama 2002). We explored the reason for this state of affairs and the potential limitations.

### 6.3.1 Case Studies

#### *6.3.1.1 Cerveza*

In November 1999, New Tokyo, a restaurant chain, released its "Cerveza" system as an open-source software compliant with general public license (GPL) to be used for ordering food supplies from stores. GPL has the following requirements: (1) the source code must be released in a manner that makes it available to anyone, (2) the released source code must be capable of alteration or redistribution by third parties, (3) the third parties must be permitted to then alter or redistribute the source code for a fee, and (4) the third parties must release the source code that they have altered, in accordance with GPL. There are numerous types of open-source software that have been released under these regulations, including Linux. However, the outstanding characteristics of this case are the facts that the software was industry-specific as well as operation-specific, and that the costs were borne by individual commercial enterprises. As of July 2004, five restaurant chains and a food wholesaler had adopted Cerveza.

#### 6.3.1.2 ORCA (On-line Receipt Computer Advantage)

The medical industry represents yet another case where specialized work applications for a specific industry have been released as open-source software. The Japan Medical Association Research Institute (JMARI), a think tank under the charge of the Japan Medical Association, has developed a system that produces receipts for the medical service fees that medical institutions claim back from payment funds on a monthly basis. This system has been named the On-line Receipt Computer Advantage (ORCA) Project, and has been released as open-source software. It went into full-scale operation in April 2002. Unlike the Cerveza system in which individual user companies are the key players, ORCA is controlled by a user organization. The goal is to introduce this system in approximately 10 000 of Japan's 100 000 medical institutions within 3-5 years.

### 6.3.2 Expectations for Greater Added Value and the Organization of Collaboration Through the Emergent Evolution of Systems

One can easily imagine the advantages of using open-source software based on those that have already been developed: reduced costs, shorter development time, as well as proven reliability based on the knowledge that the system has been used before. However, the question arises regarding the advantages for the original developer of the system. New Tokyo's primary motivating factor for developing the Cerveza open-source code was that other users would add new functions and maintain the system. In reality, other user companies have added functions to Cerveza and made improvements at their own expense, and the results of these system refinements are being shared amongst user companies. Also, JMARI, the driving force behind ORCA, is hoping that information system vendors around the country will develop subsystems with a regional focus to cater to the independent medical systems of local authorities and other related systems for electronic medical records and so on, and that this will lead to the advancement of the system as a whole. Original system developers hand over their open-source systems to several users and system vendors for improvement and development, with the expectation that the added value generated from this will exceed the initial development costs.

The problem lies in the creation of a framework for the maintenance and development of systems so that the system can undergo emergent evolution through the efforts of several users. In the case of Cerveza in particular, the largest issue will be the organization of collaborative efforts amongst companies on an equal footing so that decisions may be made regarding the type of functions to be added and the bearing of costs. The organization of collaborative efforts is in reality far more difficult than managing the technical aspects such as managing versions or parts, or the pooling of joint development know-how.

In the case of Cerveza, system improvements are basically dependent on the voluntary actions of the user companies, and there is virtually no cooperation taking place at an official level. Furthermore, support for system implementation in some cases functions on a personal level, going beyond company-level relationships. It is unclear whether this type of system would function if the number of user companies were to increase. However, this type of system, where development is left to the discretion of companies with no control measures in place and with very flexible collaborative arrangements, is set to attract attention as a new model for intercompany collaboration.

### 6.3.3 Avoiding Reliance on Information System Vendors and Reforming the Information System Industry Business Model

The secondary motivational factor for making both Cerveza and ORCA open-source software was that the developers wished to prevent their information systems from being dependent on a particular information system vendor. Systems commissioned for development, by rights, should belong to the user company that had commissioned the development. However, once system development is entrusted to a particular information system vendor, the users are forced to rely completely on that company for system maintenance and support. In the case of systems commissioned for development, the costs of providing maintenance and support usually outweigh the initial development costs.

From the perspective of those in the information system industry, the release of open-source work applications that hand over development of individual systems to customers merely serves to debase the industry's business model, which has so far allowed them to retain customers and earn profits from maintenance and support. Not surprisingly, the response of general information system vendors to the release of open-source work applications has been somewhat lukewarm. On the other hand, the company that undertook the development of Cerveza sees this as a chance for small-to-medium sized system vendors to build up business by providing maintenance and support for open-source software; while JMARI, the driving force behind ORCA, believes that this will generate new business opportunities for maintenance and support and the development of related systems. Thus, it can be said that the release of open-source software for work applications has the potential to change the business model of the information system industry.

## 6.4 Conclusion

This chapter has examined community sites that are a forum for word-of-mouth communication on products and the release of open-source software for work applications as examples of how added value is created by the intersection of consumer and company knowledge. In the given example of a community site, we saw that a new consumer-driven business model is emerging with the changes in the cost structure of communication brought about by the Internet. However, the value of these community sites does not just lie with the active participants, as was generally thought to be the case, but also with those passive participants who merely read without commenting. Moreover, in the given example in which information system user companies and groups are releasing open-source software for work applications developed at their own expense, we see that this is motivated by hopes for the emergent evolution of the system and the desire to avoid dependence on a given information system vendor. This is carried out by not keeping intellectual property under lock and key in order to enable its use and refinement by anyone. These findings have not yet been substantiated to the extent required for popularization, but we hope to accumulate further research material and establish a framework for a business model that will induce new value creation.

## References

Hagel J III, Armstrong AG (1997) Net gain. Harvard Business School Press, Boston

Isagai Y, Kokuryo J (2002) Tr@Box, Inc. 2002 (in Japanese). Keio University Keio Business School Case

Kokuryo J (1998) Economics of commercial activities in cyberspace (in Japanese). IPSJ Magazine 38:764–771

Levine R, Locke C, Searls D, Weinberger D (2000) The Cluetrain Manifesto: the end of business as usual. Perseus, Philadelphia

Ogawa M, Kokuryo J (2002) istyle Inc. (in Japanese). Keio University Keio Business School Case

Ogawa M, Sasaki Y, Tsuda H, Yoshimatsu T, Kokuryo J (2003) Silent members (SM): their communication behaviour and influence on purchases of others. Proceedings of 7th Pacific Asia Conference on Information, Adelaide, Australia, 10-13 July 2003

Takeda Y, Yoneyama S (2002) Cerveza: why did New Tokyo publicly release its company system? (in Japanese). Hitotsubashi Business Review 50:146–165

# Part III: Structural Changes

# 7 The Evolution of Organizational Structures in a Networked Society: The Case of Finland

Paul Lillrank

Department of Industrial Engineering and Management, Helsinki University of Technology, Box 5500, FIN-02015 TKK, Street address: TUAS Bldg, Otaniementie 17, Espoo, Finland, paul.lillrank@hut.fi

**Summary**

The information technology (IT) boom peaked in around 2001. Since then many innovative ventures have been forced into significant downscaling or bankruptcy. Many public programs, such as e-Government and e-Learning have remained intact, but have still been forced to accommodate new realities. The question thus is: what remains as the solid and potentially valuable core of the "new economy" and "networked society" hype? Internet-based business models have primarily led to price erosion and put increased competitive pressure on corporations, making consumers the main benefactors. Productivity increases have mainly benefited the IT-producing industries, while the potential has not necessarily been fully realized in IT-using industries. Government operations and public sector services could potentially benefit significantly from increased use of IT and networked models. However, the public sector lacks an internal innovation system and organizational frameworks for continuous improvement.

**Key Words** Information society, Demand-supply chain, Innovation systems, Administrative automation, Productivity paradox

## 7.1 Introduction

During the years around the turn of the century, the economy and society of Finland made remarkable progress. In the early 1990s, the country was hit simultaneously by a banking crisis and the collapse of Soviet Union, which almost overnight cut Finland's export earnings by 20%. During the recession, the county lost 14% of its Gross National Product (GNP), the biggest peacetime loss on record. The recovery started in 1994 and was accompanied by several policy changes. The Finnish government, based on a coalition ranging from former communists to conservatives, presented a reform agenda on which the main issue was to turn Finland into a society of knowledge and competence. Some tax and labor market reforms were initiated and an innovation system was created based on public–private cooperation and substantial investments in education and research. The most spectacular success was Nokia. The company transformed from a lumbering conglomerate to global leader in telecommunications (see Ali-Yrkkö 2001). At around the turn of the century, Finland held top positions on several international competitiveness ranking lists.

A great deal of hype was created around this success story that in many ways resembles the excitement about Japanese-style management in the 1980s. The collapse of the dot-com boom in 2000 and the ensuing economic slowdown can be interpreted as an adjustment to new technologies. The real and lasting impact of technology on society is emerging slowly and painfully. The purpose of this chapter is to discuss some of the evolutionary trends visible from a Finnish perspective. It is based on ongoing research, observations, and some official statistics.

## 7.2 Business

The fundamental ideas of the dot.com boom were that a substantial part of customer interfaces would move from stores to the Web, and that traditional demand–supply chains could be disaggregated and reconstructed, eliminating unnecessary middlemen and overcoming information-based constraints. A lot of opportunities to invent new business models were perceived. The old order, however, proved to be more resilient than expected. For example, this author was involved in a venture that aimed at creating a Web-based marketplace for print jobs, such as catalogues and stationery. While substantial savings opportunities to buyers could be demonstrated, printers were reluctant to expose themselves to the open competition that a Web market would have brought. Consequently the venture failed, resulting in the loss of substantial amounts of investors' money.

In Finland, as in many other countries, it has become clear that the significant productivity improvements registered in the second half of the 1990s primarily came from the information technology (IT)-producing industries, while the productivity development in IT-using industries has been modest. The European Union has lagged behind the United States in productivity growth in IT-using service industries during the 1990s, but the difference has become smaller in recent years. The effects of such productivity improvements have been small in Finland due to the relatively small size of IT-using service industries. The exceptions have been in banking, securities trade, and professional services.

While the ideas about strategic revolutions have turned out to be futile, a substantial amount of operational evolution has taken place. The Web and advances in software engineering have led to a significant simplification of business systems. Business applications that a decade ago would have required six man months of Cobol programming can today be put together in one afternoon using Excel spreadsheets. Cumbersome, tailor-made, and expensive order-handling systems using electronic data interfaces (EDI) have been replaced by comprehensive enterprise resource planning (ERP) systems that utilize generic software such as browsers.

For example, Jaakko Pöyry Co. Ltd, the world's leading forest industry consultancy, has developed a system called WebPub. It allows users to manage various types of project documentation is a systematic manner. The application uses generic software that allows users to view not only text and basic graphics, but also drawings made with AutoCad using ordinary browsers equipped with plug-ins. The tens of thousands of documents required in a construction project can be placed on a server, eliminating the need for paper-based, and thereby, location-dependent documents. This system has been integrated as a part of Pöyry's Virtual Mill concept, where all the documents related to the particular paper mill can be accessed from all locations, thus eliminating the need for printed materials. This allows faster response by, for example, maintenance workers, and has resulted in cost savings.

Such advances can be termed "administrative automation." In line with previous generations of factory and office automation, various administrative tasks can be managed more efficiently than before, particularly transaction management in supply chains, precision forecasting in demand-chain management, and document control for project-based business. This obviously means substantial cost savings. Simultaneously, many companies in traditional industries, such as paper and pulp and machinery, are facing significant global competition and price erosion. The substantial resources poured into technical research and development may not necessarily lead to increased pricing power. Therefore, many Finnish manufacturers, such as Kone, Wärtsilä, and Metso, have started to transform themselves from being machine suppliers to service providers. The idea is to leverage deep technical know-how from servicing the installed base. In services, various performance-related pricing schemes are possible. This, in turn, creates demand for innovations in business process configurations, in which information technology might be profitably used.

These observations have put into perspective the decades-long debate about the "productivity paradox," also known as the Solow paradox after Nobel laureate Robert Solow, who quipped that "computers are seen everywhere except in the productivity statistics." The business benefits that derive from IT and their exact measurement have long eluded researchers and practitioners. The productivity paradox is the summary name of observations that indicate on a macro level that it has been difficult to find correlations between the level of IT investments and productivity, or any financial indicators (Strassman 1991, Lucas 1999).

On a macro level, the main dependent variable is total factor productivity (TFP). Its rapid increase during the early part of the 1990s prompted a broad discussion about "the new economy." However, closer scrutiny of the data reveals that TFP growth has been rather unevenly distributed among various industries. A major part of the growth impact comes from IT-building industries, while for many IT-using sectors productivity growth has been unremarkable or even stagnant. The ensuing debate can be summarized into five arguments (Willcocks and Lester 1999):

1. The statistics measure the wrong things. A great deal of the impact of IT goes into services and administrative processes, where productivity impacts are hard to measure. Price competition erodes financial benefits of investments. Poor performers not investing in IT go out of business and thereby disappear from the statistics. The primary benefits go to customers in the form of qualitative improvements of products and services.
2. IT is still a rather minor part of the general capital stock and therefore cannot have a dramatic impact on productivity. The rate of IT investments out of total investments, however, is growing rapidly being around 40% in the mid 1990s. The numbers are difficult to interpret because a large part of IT investments are replenishments of old equipment, thereby the growth of the entire capital stock may be slower.
3. IT will have an impact on productivity as soon as the institutional environment is adapted to the new technology. This argument draws on historical parallels with other technologies such as steam engines or electric motors, for which it took several decades to have a serious impact on the economy.
4. IT has an impact on productivity, but only for pioneers who turn it into a strategic advantage by integrating it to other competitive advantages, such as a highly skilled workforce and modern incentive systems. Case studies about industry leaders integrating IT with new organizational arrangements tend to support this view.
5. IT really does not have much productivity potential in the first place. Compared with other innovations, such as electricity, internal combustion engines, or

petrochemicals, IT is basically only "a little more of the same," offering small improvements in administrative and control functions. Further increases in memories and processor speed have very marginal impacts.

While the debate is far from over, a few conclusions on a corporate level can be drawn. Straightforward IT applications in factory and office automation are easy to imitate. As all or most competitors in an industry have made the basic investments, competition will equalize the competitive advantages acquired by IT. The winners will be customers, who will see the benefit of improved price/quality ratios in many deliverables. In this respect the question is not about investing for profit, but for survival. Basic IT investments take the same status as telephones, copying machines, and office furniture: they are necessary infrastructure appliances that do not warrant separate payback calculations.

On the enterprise level, sustainable competitive advantage can be derived from company-specific or business process-specific IT applications that are hard to imitate. These require, in addition to IT hardware and software, significant amounts of organizational and managerial skills to initiate, plan, and execute. Networked structures involving social capital, such as mutual trust, are seen as having a high capability for sustainable competitive advantage. However, it remains unclear to what extent these kinds of structures can respond to the challenge of developing not only innovative gadgets, but also systems that make business use of them.

## 7.3 Innovation System

In the depths of the early 1990s recession, the Finnish government proclaimed a national vision based on knowledge and competence. A substantial amount of public funds was channeled through public bodies with the purpose of supporting corporate ventures, research and development, and new business development. Frameworks were designed to enhance business–academia cooperation. During the economic slowdown after the dot-com boom, developing new innovations and enhancing competitiveness in the high-technology sector have remained high on the Finnish government's agenda, despite drastic reductions in the amount of new technology-based ventures. For instance, the government under Prime Minister Vanhanen has launched an information society program (www.infosoc.fi) as one of four cross-ministry programs. It is officially aimed at improving competitiveness, productivity, and the quality of life of citizens. The program consists of nearly 100 development projects divided into eight areas of focus, covering nearly all areas of application for new information technology. The areas of focus, for instance, include government IT services, as well as telecommunication infrastructure and digital television.

### 7.3.1 Financiers of Innovations

The following major public bodies constitute the Finnish innovation system:

The Academy of Finland is a major financier of academic research with a focus on basic and precompetitive research. It is under the jurisdiction of the Ministry of Education and covers all major scientific sectors. Its total grant budget was close to 200 million euros in 2004. In 2003, the Academy of Finland was responsible for distributing 13% of the total research funding provided by the Finnish government (see http://www.aka.fi).

TEKES is an organization under the Ministry of Trade and Industry and is focused on technology development and applied research. Its primary objective is to promote the competitiveness of Finnish industry and the service sector by technological means. Activities should lead to more diversified production structures, increased production and exports, and create a foundation for employment and societal well-being. TEKES's main functions are defined as framing and preparation of national technology policy; preparing, financing, and coordination of national technology programs; financing applied technical research and risk-intensive industrial research and development projects; financing and coordination of international and regional technological cooperation; and providing small and medium-sized enterprises (SMEs) with advisory services in technology transfer and exploitation of new technologies. TEKES works primarily by providing selective project funding and expert services to research projects run by companies, research institutes, and universities. TEKES focuses on research that can be commercialized and promotes sustainable competitiveness. Although they aid in the development and finalization of ideas, TEKES does not claim any intellectual property rights to the projects that they assist. TEKES distributed approximately 400 million euros in 2004, to close to 2000 different projects (see http://www.tekes.fi).

SITRA, the Finnish National Fund for Research and Development, is an independent public foundation operating under the responsibility of the Finnish Parliament. SITRA invests in technology firms and venture capital funds, both in Finland and abroad, and finances and implements research, training, and innovative projects of significant national importance. Funding is given to companies in their seed or start-up stages of development or for SMEs in their early stages of development seeking to expand internationally. Special focus areas are technology-intensive or health care companies. SITRA has also served as a think tank for the government's information society initiatives, doing research on various topics related to its key areas of interest, such as innovations and new technologies and their effects (see http://www.sitra.fi).

TIEKE is the Finnish Information Society Development Center. Its main functions include initiation and coordination of selected development projects within the IT industry; generation of an IT cluster together with domestic and international partners, thus promoting cooperation and standardization at all levels of IT industry; development of trade procedures in both domestic and international cooperation as a major Finnish player in that arena; and participation in the standardization of information and communications technologies at national, European, and international levels. TIEKE also offers IT training to citizens and advises the Finnish government on issues related to the information society, such as digital television, healthcare, and e-Learning. It is a considerably smaller organization with an annual budget of around 2 million euros, and, unlike the others presented here, does not distribute actual funding for research and development. Despite this, TIEKE plays a key networking role in connecting various players in the development of the Finnish information society (see http://www.tieke.fi).

In Finland the relation between industry and academia has traditionally been smooth and intimate. In particular, TEKES has played a significant role in building these relationships. A major part of research funding is structured as follows. Any academic research group with an idea can apply for funding for initial research or feasibility studies. TEKES also manages several large research programs in various fields. Companies and universities are invited to send proposals for various projects that fit into the program area. Academic researchers are obliged to find industry partners to fund 40% to 60% of total project expenses. Thus, research teams often include both academics and company staff.

The innovation system has produced substantial amounts of academia–industry joint programs and projects. Another aspect of the innovation system is the promotion of innovative use of IT, especially among SMEs.

### 7.3.2 Policy Mechanisms

The various policy mechanisms used to promote the use of information technology are based on certain assumptions that lead to specific schemes, which may or may not produce intended results. In a discussion paper presented by the Benchmarking Coordination Office of the European Commission DG III, the basic policy approaches were summarized as follows.

1. Standardization of various exchange regulations, communication protocol, and

the legal environment is the traditional role of government. The assumption is that various emerging standards struggling for dominance in the marketplace may lead to suboptimal allocation and delays in the introduction of new technologies and practices. While there are cases, such as in the banking sector, in which private agents have swiftly created and implemented de facto standards, there may be areas in which government initiatives are beneficial.

2. Network building. The underlying assumption is that SMEs in particular learn most rapidly from each other. The role of public bodies is to facilitate networking between SMEs on a regional or industry basis, for example, by providing the initial investment for a communication platform. This would allow mutual learning, coevolution, and rapid dissemination of best practices.
3. General network building. The assumption is that SMEs have problems in linking to various knowledge providers and developers outside their own industries or regions. Therefore, networks, including universities, research institutes, testing and accreditation bodies, and other knowledge providers are established.
4. Cluster support. The assumption is that technical and managerial development is led by larger enterprises that serve as "industrial locomotives" in regional or industrial clusters. They will pull their suppliers into adapting advanced methods in the course of regular transactions and occasionally provide technical support, for example, in adapting standardized order and invoicing systems, extranets, and integrated logistics. By way of example, they may have a positive impact even beyond their direct business partners.
5. General awareness building. The assumption is that SMEs generally are not sufficiently aware of the latest developments and do not have direct links to academia or other sources of innovation. Therefore, public bodies should provide information about the potential of information technology.
6. Specific awareness building. The assumption is that although SMEs may, on a general level, be aware of the potential of IT, they do not have the resources to develop insights relevant to their particular line of business. IT can be used in different industries in different ways. Specific awareness building concentrates

on some sectors or regions and supply relevant and targeted information, often best-case descriptions that are hoped to arouse interest among SMEs.

7. Consulting support. The assumption is that although SMEs may be aware of the potential of IT in their business, they lack the resources to conduct the essential analysis and planning tasks required for implementation. Therefore, specific consulting support is provided on a subsidized basis.
8. Specific financial support services. The assumption is that SMEs have problems in raising the necessary capital for IT investments. Therefore, targeted financial support schemes are offered.

Various support schemes have been devised by several EU member countries. However, in recent follow-up studies, it has been rather difficult to prove that these schemes would have significant positive effects.

## 7.4 Public Sector

e-Government can be seen as having three different parts: the administrative, the political, and the service-production parts. In a narrow sense, e-Government means that various administrative government functions and administrative services for citizens and businesses are conducted with the help of networked IT. A main part of governments' tasks are collecting, processing, analyzing, distributing, and acting upon information in various forms. Much of the administrative products of government, such as decisions, statements, permissions, licenses, and certificates, are rather easily converted into digital form. Such administrative aspects of government can be made increasingly efficient with the aid of office automation, databases, and networks.

The various relations between public administration (A), citizens or customers (C), and businesses (B) can be summarized as:

- A2A: intra-administrative transactions between various agencies, exchange of data and information, such as calculations of GNP, census data collection, and various registers

- A2B: administration to businesses, such as regulations, permits, and value added tax (VAT)
- A2C: services to, and transactions with private citizens, such as personal documents and building permits

In a broader sense, e-Government refers to changes in political processes that become possible with networked IT. In its most radical form this would mean direct democracy, where citizens could cast their vote over the Internet on various political issues and thereby take over some of the functions of elected bodies. This, however, has so far not been seriously implemented. A more realistic aspect is the possibility of increased transparency of various political processes. Policy debates, position papers, white papers, committee work, draft legislation, and the final decision-making processes could be made available to citizens. Discussion forums can be established for public debate.

Transparent government can be divided into two parts. The first is the transparency of administration. The traditional Nordic transparency principles state "official records shall be in the public domain." Authorities are required to publish information proactively. There is a shift from "supplied when asked" to active information policy. Several joint information services are available for citizens. For instance, a citizen's guide to public on-line services has been jointly produced by the Ministry of the Interior and the Ministry of Finance (www.asiontiopas.fi). There is also a common portal offering links to all Internet services offered by the public sector (suomi.fi). In addition, a general directory of all public bodies and their employees has also been made available (www.julha.fi).

The second point in transparent government is the openness of policy processes from ideas to completed legislation. Preparatory work in ministries can be viewed in a project database (www.hare.vn.fi). All ministries publish their reports and proposals on their websites. Government proposals to parliament can be read at the Council of State website (www.vn.fi). Discussions and decisions in parliament are available at www.eduskunta.fi, and published laws can be viewed at www.finlex.fi.

In addition to services providing general public information to citizens, the

Finnish government has also begun offering access to services requiring authentication. Examples of such services are reviewing of an individual's social security benefits, accessing tax records, and employment services. These services can be accessed using an electronic ID card, which have not yet achieved much popularity in Finland, or by using the Internet banking IDs of most major Finnish banks. There also exists a centralized service providing forms required for various government agencies (lomake.fi). Currently, a majority of the forms still need to be printed out and returned using conventional means, but the share of forms that can be filled out and returned electronically is continuously growing. An example of such a service is the ability to report minor offences, such as theft, to the police using an electronic form, eliminating the need to visit a police station.

An aspect that follows on from the transparency of government is the question of public service production, including education, welfare, healthcare, safety, security, and various infrastructure services. In service products, such as healthcare, the patient is simultaneously subject to a service intervention and is the benefactor of the service. In administrative products or processes, such as in the case of the highway police apprehending a speeding motorist, the offender is subject to an administrative act, but the main benefactor is the public that benefits from safe road traffic. Public service production can benefit from IT-enhanced service processes in a similar vein to private businesses. It is said, however, that when it comes to service-process management in Finland, the public sector lags the private by about 30 years. In health and welfare services, a general perception is that the actual services are of high quality; however, the coordination and administration of other than the simplest tasks is a problem. This is especially so in complex cases in which two or more sectors, such as school, health, and welfare professionals are involved. Inefficiencies due to poor coordination and unclear priorities lead to misalignment, long waiting times, and wasted effort. Unofficial estimates, publicly expressed by high-ranking Ministry of Trade and Industry officials, claim that overall cost improvement of around 20% could be achieved with better management and coordination, while maintaining or even improving quality levels. To achieve this potential requires more than installation and use of standard IT solutions.

One example of improving the efficiency of public services is the introduction of various administrative automation systems in the healthcare sector. The most prominent one of these is the implementation of a nationwide system for handling electronic patient records and referral-treatment feedback. The nationwide system is intended to consist of separate systems in each hospital district, each possibly provided by a different software developer, communicating through standardized protocols. The system is currently being implemented and, according to plans, will be ready in 2007. The development of the nationwide system is financed partly by the government and partly by the hospital districts themselves as part of normal IT system upgrades. The largest hospital districts have already adopted their own systems for handling electronic patient records, as well as related data such as X-ray and other imaging pictures and electronic prescriptions. The majority of these systems have been produced by domestic software developers, and this has also aided in reaching an agreement on the standardized communication protocol. Despite this, modifying the existing systems to be compatible will still pose challenges in the future.

The parties to the nationwide project, including software developers, hospital districts, and an international standards organization specializing in healthcare IT protocols (HL7), with coordination by the Ministry of Social Affairs and Health, have agreed on a joint protocol that will form the basis of the system in the future. There are also additional systems being implemented within hospital districts, such as electronic prescriptions, that are not yet covered in the nationwide standard, but that will integrate with the electronic patient information systems. Most of the benefits from implementing such administrative automation systems have not yet been realized, but projected benefits include cost savings and improved speed of treatment. One proposed use for the electronic patient records is consultations, for example, a doctor working at a healthcare center can request a consultation from a specialist at a hospital, who will have immediate access to all relevant patient data. This will often eliminate the need for the patient to actually visit the hospital, resulting in significant cost savings.

## 7.5 Society

Castells and Himanen have conceptualized three basic models for the information society: the Silicon Valley, the Singaporean, and the Finnish models (Castells and Himanen 2002; see also Himanen 2001). These are differentiated between by two variables: the openness vs. closeness on society; and the activity vs. inactivity of government in various social issues, including promotion of information society initiatives. The US model is, obviously for an open society with reasonably little government involvement, where market forces are assumed to do the job. Singapore, on the other hand, has a more closed society and an activist government, which have led to several top-down authoritarian initiatives. Finland, according to Castells and Himanen, differs from the above-mentioned two in the respect that the society is open while government is active.

Castells and Himanen drive the thesis that government activism, in providing infrastructure, education, and publicly funded innovation systems, is necessary for developing and implementing the expected benefits from information technology. The validity of this thesis obviously carries significant policy implications. While the activist government thesis looks attractive, the ongoing recession has prompted critical voices within Finland. It appears that most of the evidence given to support the benefits of the innovation system is related to activities. It is obvious that an activist government creates a lot of activities, which in turn have an impact on perceptions, such as rankings in competitiveness. There are, however, increasing doubts whether the billions poured into research by TEKES have really succeeded in creating sellable products, sales revenue, or entirely new industrial branches. A second doubt is related to the payback of education. Currently, over 70% of school leavers go on to tertiary studies at universities, polytechnics or vocational programmes. The number of new PhD degrees has tripled in a decade. The verdict is still out whether this has resulted mostly in an inflation of job titles and qualifications and not much more. In summary, the fundamental question relates to whether government-sponsored activities really lead to the creation of new value, which leads to new utilities to citizens and to new products and services that can withstand the general price erosion caused by global competition.

There has also been criticism of the Finnish government's information society policy being too focused on technological development and neglecting other aspects of societal change related to the development of the information society. Castells perceives the information society as a result of a holistic change in the entire society, rather than just technology. Other areas in need of special attention include responding to various global trends, the increasing role of creativity as a source of competitive advantage and the means to support it, the role of the government in the provision of public services, tax, and immigration policies as well as ensuring well-being at work. The responses to global trends such as the new global division of labor are common topics of discussion in general policymaking, but their links to the information society policy have been largely ignored, at least as part of public policy. The information society has been presented as a cure to the problems caused by global competition or the change in population age structure, but more in-depth analysis has been rare. This limits the opportunities to evaluate the possible larger-scale impacts that Finland's information society policies have had so far.

## 7.6 Conclusion

The most significant impact of information technology on society appears to come through two channels. First, technology enables new labor-saving and energy-saving devices, most recently administration automation, to reduce costs and improve quality in various sectors. The second impact comes through technology-enabled networking, which brings resources, people, knowledge, and ideas together in new ways. Networks are based on social structures, relations, and capital, which are slow to emerge. New types of social arrangements can be observed in embryonic form in the supply chain and research and development management systems in some high-technology industries. Since the IT hype that occurred at the turn of the century, the general mood in Finland, amplified by the prolonged recession, appears to be have been tuned to the cost-saving aspects of information technology. Especially the private sector has experienced a period of relatively small advancements in the use of IT due to a lack of investment, but at the same time the public sector has continued developing its services in accordance with public policy. With the economy now showing some signs of recovery, a renewed interest in IT in the private sector is also becoming apparent. Despite this, the full effects of the recent advancements in information technology have not yet become visible.

## References

Ali-Yrkkö J (2001) Nokia's network—gaining competitiveness from co-operation. Taloustieto, Helsinki

Castells M, Himanen P (2002) The information society and the welfare state: the Finnish model. Oxford University Press, Oxford

Himanen P (2001) The hacker ethic and the spirit of the information age. Random House, London

Lucas HC Jr (1999) Information technology and the productivity paradox. Assessing the value of investing in IT. Oxford University Press, New York

Strassmann PA (1991) The business value of computers. An executive's guide (revised and updated edn). Information Economics, New Canaan, CT

Willcocks LP, Lester S (eds) (1999) Beyond the IT productivity paradox. Wiley, New York

# 8 Research and Development Partnerships Bridged by the Intermediate Domain: A Research Provider in the Semiconductor Industry

Hideyuki Tanaka

Graduate School of Interdisciplinary Informatics, The University of Tokyo, 7-3-1, Hongo, Bunkyo-ku, Tokyo 113-0033, Japan, tanaka@iii.u-tokyo.ac.jp

**Summary**

There are two major types of partnerships in research and development: one centered on relationships among businesses and the other centered on cooperation between industry and academia. This chapter shows how these two types of partnership become merged in the form of partnerships in the intermediate domain, which lies between the private domain of corporate business activity and the public domain of university research. This study applies this framework to the Interuniversity MicroElectronics Center (IMEC), a research and development institute in the semiconductor industry, to explain how a research provider in the intermediate domain functions independently as a node for research and development partnerships.

**Key Words** Research and development, Cooperation between industry and academia, Innovation, Intermediate domain, Semiconductor industry

## 8.1 Research and Development Partnerships and the Intermediate Domain

### 8.1.1 Current State of Research and Development Partnerships

This chapter begins with a brief examination of the current state of research and development (R&D) partnerships. Since the 1980s, R&D partnerships[1] have been widely used in corporate R&D activities. Generally reviewing the R&D activities of American businesses in the twentieth century, Hounshell (1996) revealed that there was a significant rise in the number of firms engaged in joint R&D projects in the period from the 1980s to the early 1990s. Moreover, R&D partnerships now transcend national boundaries, instead of being confined within them (Mowery 2003; Teece 1996; Edler et al. 2002). These tendencies have been empirically verified by research not only on US firms, but also on European and Japanese businesses (Roberts 2001; Edler et al. 2002).

There are basically four factors behind this spread of R&D partnerships. Specifically, these are: (1) cost increases, (2) a rise in risk, (3) a growing need for flexibility, and (4) globalization in the field of research and development.[2]

First, cost increases may for instance refer to a situation in which it is increasingly difficult for individual companies to withstand the dramatically rising costs of new product development (Mowery 2003; Teece 1996). In the semiconductor industry, which is the focus of this chapter, it has been pointed out that it is now difficult for individual businesses to cover these costs as a result of

[1] "R&D partnership" broadly refers to relationships based on innovation. They can take various forms on the condition that the participants share the resources for reaching common R&D goals. They range from joint ventures based on investments to research alliances and consortiums based on contracts. Participating entities are not limited to profit-making businesses. R&D partnerships can be formed among different types of entities including government research institutions and universities (Hagedoorn 2002; Link and Vonortas 2000).

[2] An additional factor applicable to American firms is the deregulation of joint research brought about by the National Cooperative Research Act in 1984 (Mowery 2003; Teece 1996).

the progress in technology (Ministry of Economy, Trade and Industry 2002, Nishi 2003).

Second, R&D activities for advanced technologies involve high uncertainty and therefore a high risk as well (Doz and Hamel 1998). This high risk along with the growing R&D costs is becoming increasingly burdensome for individual companies and generates a need for R&D partnerships.

Third, there is a growing need for strategic flexibility to keep up with dramatic changes in technology and markets (Sanchez 1995). The options provided by available resources must be secured to bolster flexibility. In view of mounting R&D costs and risks, it is becoming more appropriate to establish strategic partnerships than to persist with independent development (Teece 1992; Hagedoorn 2002; Link and Vonortas 2000). A recent empirical study (Hagedoorn 2002) confirmed that flexible partnerships are now created with the main stream of R&D partnerships shifting from joint ventures based on investment to partnerships based on contracts without investment. The trend for flexibility is particularly prominent in high-tech industries.

Fourth, with respect to the globalization of R&D activities, Mowery (2003) and Teece (1996) reported that cutting-edge R&D activities are now taking place around the world. Empirical research into direct foreign investment (Florida 1997; Edler et al. 2002) confirmed that direct investment in research and development, which was conventionally aimed at localizing products (Vernon 1966), is now made for the purpose of actively using the most advanced overseas research and technologies. In addition, a rise in the percentage of joint research papers produced by multinational teams of researchers, from 14% in 1986 to 31% in 1999, reflects this R&D globalization (OECD 2002). In the semiconductor industry covered by this chapter, American firms alone could not carry out the R&D project on extra ultraviolet (EUV) technology, which is one of the nanoscale semiconductor manufacturing technologies. Although the project used a federal government scheme known as the Cooperative Research and Development Agreement (CRADA), which was an agreement on joint research between federal government research institutions and private businesses, it had no choice but to invite the participation of a European firm, namely ASML (Leopold and Lammers 2000). The semiconductor industry has already entered a phase in which research and development can no longer be made by companies in a single country alone (Mowery 2003).

So what is the potential domain of R&D partnerships currently being pursued of the type described above? This chapter examines partnerships among private businesses separately from partnerships between businesses and academia.

### 8.1.2 Research and Development Partnerships Among Businesses

Earlier studies have shown that R&D partnerships within the private sector are seen in precompetitive domains (Fusfeld and Haklisch 1985; Sandholtz 1992; Gomes-Casseres 1996).

The definition of precompetitive domains is premised on the classification of R&D domains into three as in Fig. 1: "noncompetitive," "precompetitive," and "competitive." In precompetitive domains, R&D cannot be carried out by a single company alone even though it involves a technological foundation necessary for ensuring corporate competitiveness. Precompetitive domains differ from noncompetitive domains in that they are strongly associated with corporate competitiveness and they offer each enterprise an incentive to invest in R&D activities. However, R&D achievements in these areas are likely to spread and provide benefits to other companies as well. There is little possibility that a single company can enjoy an exclusive advantage, and if separate firms made their own efforts their investment would be too small when compared with the social optimum (Arrow 1962). On the other hand, R&D partnerships in precompetitive domains motivate businesses to make a greater R&D investment than they would for separate R&D activities (D'Aspremont and Jacquemin 1988). This proves that R&D partnerships in precompetitive domains do make sense. Since the 1980s, we have seen plenty of R&D partnerships in precompetitive areas among companies within the same country or region. This is partly because of policy efforts, including revisions of antitrust legislation made by national governments in response to the growing importance of national competitiveness.

**Fig. 1.** Three levels of research and development

| |
|---|
| Competitive |
| Precompetitive |
| Noncompetitive |

This chapter now redefines precompetitive research from the perspective of corporate assets. Bressand (1991) employed the strategic and organizational distinctions as two axes to draw a map of corporate assets (see Fig. 2). In the past, it was simple to distinguish what was internal to a company from what was external to it. It was possible for companies to strategically control the use of accessible assets inside their corporate organization while it was impossible to exercise strategic control over inaccessible resources outside their structures. In other words, there used to be a general correspondence between "organizational" control over access to corporate resources and "strategic" control over the use of those resources.

**Fig. 2.** Map of corporate assets. Developed with reference to Bressand (1991)

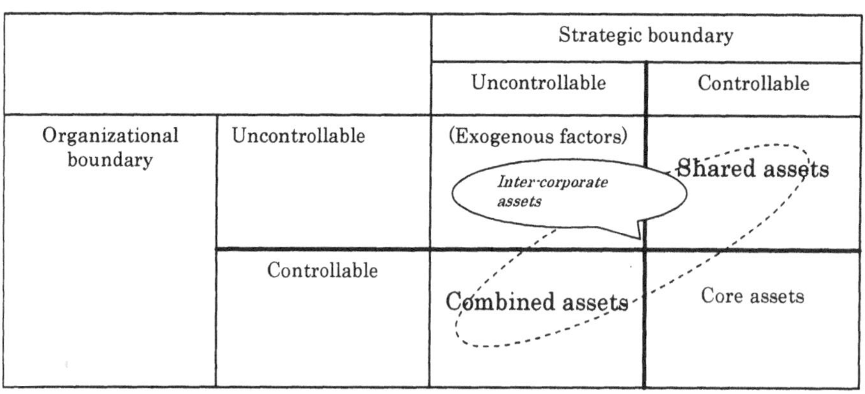

However, the growing importance of intercorporate networks has given rise to some corporate assets where there is no longer any correspondence between the two forms of control. One example is shared assets, which are positioned outside the organizational boundaries of a company but can be strategically controlled to a certain extent. Another is combined assets, which are under the organizational control of a company but can also be put under the strategic control of different economic entities.

In accordance with this mapping, precompetitive R&D partnerships are positioned as shared or combined assets. In a case in which each participating company conducts joint research within its own premises, individual research activities can be under organizational control because they are conducted within the organizational borders of the respective firms. However, strategic control is not necessarily possible as the research strategy is also subject to the strategic control of other partners. Joint research of this type is classified as a combined asset. If joint research is conducted in an independent research institute, it is seen as a shared asset from the viewpoint of participating companies. It can still be under their strategic control even though it is performed outside their organizational boundaries.

Sudoh (1995) stresses that strategic use of assets in the intercorporate domains such as shared and combined assets is significant for the acquisition of corporate competitiveness in the trend toward corporate networks. The rising importance of R&D partnerships in precompetitive domains is evident from the viewpoint of intercorporate assets.

### 8.1.3 Research and Development Partnerships Between Industry and Academia

In the United States, collaboration between businesses and academia is extremely widespread, backed by a number of policies introduced in the 1970s and onward. In Japan as well, industrial–academic cooperation has become more prominent in recent years. It is possible to suggest two possible reasons for the rise of industry–university partnerships. The first is a problem with basic research conducted by businesses. In the late 1960s, it was acknowledged in the United States that few cases of corporate basic research distinct from development had successfully led to corporate growth (Hounshell 1996). Basic research provides a foundation for practical application (Bush 1945), but it has been confirmed that it is important for businesses to take advantage of research achievements produced by universities rather than carrying out such research on their own. The second is the issue concerning the role of universities in equipping a nation for industrial competitiveness. The growth in university–business collaboration is linked to the idea that the results of basic research produced by universities must be actively used to generate industrial competitiveness (Council for Science and Technology Policy 2002).

Research and development cannot be simply split into two, namely basic research and applied research and development. Stokes (1997) plotted a matrix of research and development with considerations of use and fundamental understanding forming the two axes. What has traditionally been called "basic research" is divided into two: what is done in consideration of use and what is not. What has conventionally been called "applied research" is also split into two: that which requires fundamental understanding and that which does not. Fig. 3 clearly demonstrates that there are three types of research and development, namely (1) pure basic research, (2) use-inspired basic research ("Pasteur's Quadrant"), and (3) pure applied research. Research and development at the precompetitive stage mentioned in the previous section falls into the category of basic research with consideration of use in the sense that it is linked to corporate competitiveness. On the other hand, it is closely related to fundamental understanding as its outcome spreads to other companies. Hence, the precompetitive research and development to be conducted through R&D partnerships corresponds to use-inspired basic research.

**Fig. 3.** Quadrant model of scientific research

| | | Considerations for use? | |
|---|---|---|---|
| | | No | Yes |
| Quest for fundamental understanding? | Yes | Pure basic research (Bohr) | **Use-inspired basic research (Pasteur)** |
| | No | | Pure applied research (Edison) |

Source: Strokes (1997)

Universities are expected to make the results of their research widely available to society, and their activities basically belong to the public domain. In contrast, private businesses are expected to pursue their own profits through markets and their activities fundamentally belong to the private domain. With its achievements widely owned by the public, pure basic research has a noncompetitive status and is carried out by universities in the public domain. With its achievements directly connected to products and services, pure applied research and development is competitive and is carried out by private companies in the private sector. So how should the precompetitive research and development in Pasteur's Quadrant be done? Partnerships are established between universities that focus on fundamental understanding and private enterprises that focus on use. In other words, collaboration between industry and academia is a tool for research and development in Pasteur's Quadrant.

However, this type of collaboration is not always confined to the implementation of joint research and the use of university-originated patents. As seen in Fig. 4, empirical analysis of American businesses (Cohen and Levinthal 1990; Cohen et al. 2002) confirms that they attach most importance to published documents in their use of the results of research carried out by universities, followed by conferences or meetings, and informal interactions. This reveals that they put little emphasis on joint research, contracted research and patents. It is not necessarily appropriate to seek a solution to corporate product development from the outcomes of research conducted by universities. Rather, what is increasingly important is to build close ties between university researchers undertaking basic research and corporate researchers doing applied research (Rosenberg and Nelson 1994).

**Fig. 4.** Channels of research with universities appreciated by businesses

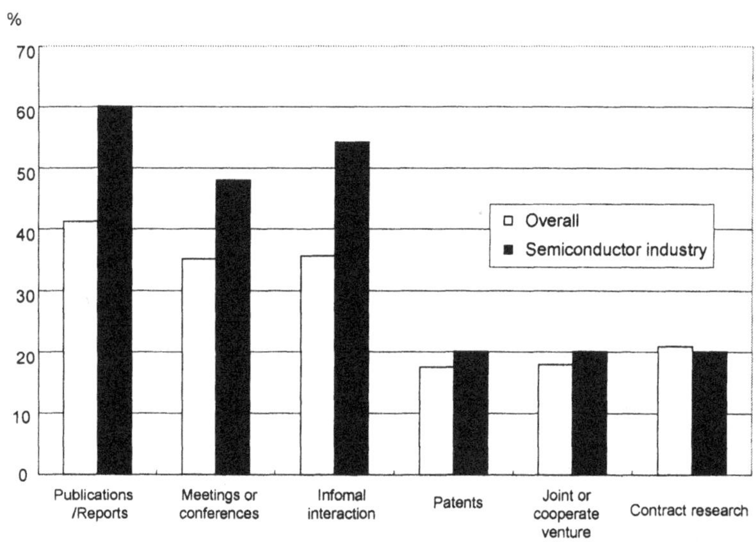


Source: prepared on the basis of data from Cohen and Levinthal (2000) and Cohen et al. (2002)

To incorporate the results of basic research at universities as outsiders, it is essential for private companies to develop an absorptive capacity (Cohen and Levinthal 1990). To do this, they must conduct basic research to a certain degree within their own organizations (Freeman 1991). But in the face of fierce international competition, private enterprises are shifting their R&D funds from basic research to applied research (Rosenbloom and Spencer 1996; OECD 2002). If this trend continues, it will be difficult for them to absorb academic research results. The establishment of close relationships between businesses and universities is dependent on research in Pasteur's Quadrant that lies between pure basic research and pure applied research and development.

### 8.1.4 The Compound Nature of Research and Development Relationships

The precompetitive area is important not only to R&D partnerships among enterprises but also to those among industry, academia, and government. In this area, the fruits are neither completely under exclusive ownership nor completely in the public domain. Given that it is between the domain of exclusive ownership, which is the private domain, and the domain of public ownership, which is the public domain, it is labeled the "intermediate domain." Table 1 provides a summary of the discussions in Sects. 8.1.2 and 8.1.3. Precompetitive research and development forms intercorporate partnerships that serve as the intercorporate assets of private businesses. At the same time, it is also research and development that links pure applied research and development with pure basic research to form industry–academia partnerships.

**Table 1.** Three domains of research and development (R&D)

| Domain | Players | Competition level | Assets | R&D |
|---|---|---|---|---|
| Private domain | Business | Competitive | Core assets | Pure applied R&D |
| Intermediate domain | (Partnerships) | Precompetitive | Intercorporate assets | Use-inspired basic research |
| Public domain | Academia | Noncompetitive | Public assets | Pure basic research |

To put it another way, R&D partnerships in the intermediate domain have two aspects:

1. A trait of horizontal partnerships that link different businesses within the private sector. Related companies cooperate in research and development, the results of which are likely to propagate to other businesses because of their basic nature.

2. Vertical partnerships that connect research and development in the private domain with that in the public domain. Applied research and new product development cannot easily be carried out in isolation from pure basic research, but private firms now have little energy to spare for pure basic research. This is where business–university collaboration gains its importance.

As illustrated in Fig. 5, intermediate-domain R&D partnerships in pursuit of fundamental understanding provide a combination of horizontal and vertical links.

**Fig. 5.** Horizontal and vertical research and development partnerships connected by the intermediate domain

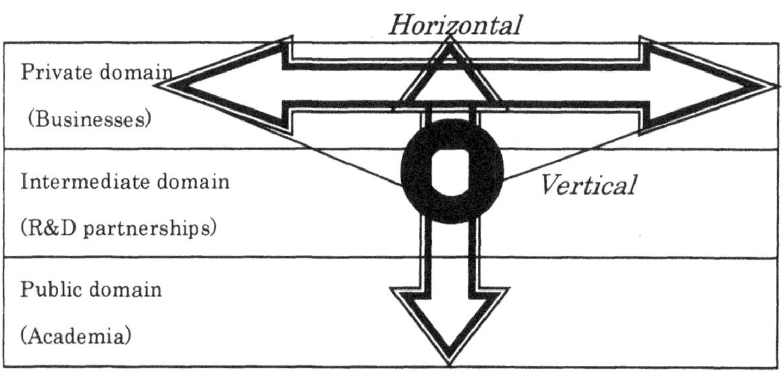


Here, a question arises: should we expect there to be any kind of independent entity in the intermediate domain corresponding to the state of enterprises in the private domain and universities in the public domain? With respect to horizontal partnerships, as mentioned in Sect. 8.1.1, joint research is being transformed from fixed projects involving expenditure to ad hoc ones based on contracts, reflecting the demand for greater flexibility. An intermediate domain could be formed without the existence of any independent entity simply by relying on such contractual relationships. In terms of vertical partnerships, it is possible to establish a connection between the public and private domains when enterprises offer funds to universities.

For horizontal partnerships, ad hoc partnerships may not be sufficient for the smooth maintenance and expansion of relationships created as intercorporate assets in the intermediate domain. It is said that success in partnerships rests on accumulation (Gomes-Casseres 1996). If ad hoc relationships are retained as they are, participating companies will not be able to accumulate a common experience and it will be difficult to create any shared foundation. In response to the need for global and flexible partnerships, it is crucial to develop an open system that enables participating firms to have a shared context.

As far as vertical partnerships are concerned, some enterprises have been urged to take a new technological approach based on different pure basic research. An example is the semiconductor industry facing the challenge of nanoscale technology reviewed in the next section. Meanwhile, positioned as they are in the private sector, these enterprises are also obliged to put further emphasis on applied research and development. Given an enormous discrepancy between the private and public domains, it is important for the private sector to develop the capability to absorb the results of pure basic research.

While it is becoming harder to construct either horizontal or vertical R&D partnerships, the semiconductor industry is demonstrating a solution to this problem by installing an independent organization in the intermediate domain. The following section applies the framework of R&D partnerships to the Interuniversity MicroElectronics Center (IMEC), which is an institution for research and development in the semiconductor field set up to examine the potential of independent entities in the intermediate domain.

## 8.2 The Semiconductor Industry and IMEC

### 8.2.1 The Semiconductor Industry Facing the Challenge of Nanoscale Technology

It became clear at the end of the 1990s that the semiconductor industry would hit a brick wall with miniaturization technologies at the nanometer level. Semiconductor technology must attain further miniaturization in order to create an environment for ubiquitous computing in which people can use computers without being conscious of them. For instance, it is necessary to embed the entire system including software on a single chip (called System-on-Chip or SoC). Extension of existing technology based on microelectronics can no longer support nanoscale downsizing. The 1999 Edition of the International Technology Roadmap for Semiconductors (ITRS) drew international public attention to this technical limitation using the expression "Red Brick Wall." More fundamental questions need to be answered before this hurdle can be cleared, and this will necessitate a revision of materials, structures, and the basic format (Yoshimi and Fukushima 2002). In the product architecture developed before the early 1990s, any innovation within an existing module would affect other modules. Therefore, all processes ranging from design to manufacture must be taken into consideration when innovating. There should be a horizontal R&D partnership with the participation of related companies like the former SEMATECH, which successfully accomplished the standardization of the present semiconductor industry.

From the perspective of academic research, research and development involving nanoscale technologies can no longer be easily covered by microelectronics alone, although this has been a major academic area in research and development on semiconductors. Microelectronics is characterized by a top-down approach in the microfabrication of objects, but it also needs a bottom-up approach for assembling atoms and molecules into nanoscale substances. This indicates the necessity of performing studies in other academic areas such as biology and atomic manipulation. Fig. 6 shows that semiconductor miniaturization technologies will reach a point at which the top-down approach and the bottom-up approach intersect, sometime around 2010. To ensure the effective advancement of innovation, it is not enough to depend only on conventional microelectronics. Basic research results in other scholastic fields must also be introduced. Vertical R&D partnerships are additionally required for the purpose of combining pure basic research with applied research and development.

**Fig. 6.** Semiconductor miniaturization technologies and nanotechnology

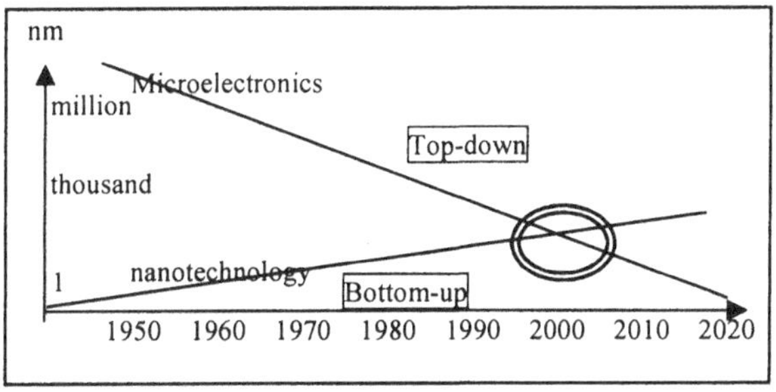


Source: prepared on the basis of IMEC materials

What measures are being taken in response to a situation that necessitates not merely horizontal partnerships among businesses, but also vertical ones that unite the private domain with the public one, as illustrated in Fig. 5 In the semiconductor industry, a brand-new type of player that differs from enterprises or universities is coming into being in the intermediate domain to bridge the gap between the two sectors to establish partnerships between them. The section that follows looks at IMEC, which is a research and development institution based in Belgium and is an example of such an organization.

### 8.2.2 Research and Development Partnerships Constructed by IMEC

The Interuniversity MicroElectronics Center, or IMEC, is a nonprofit research organization based in the Flemish city of Leuven in Belgium. Established in 1984, it has a total of 1272 staff members, including those dispatched from private companies.[3] Dealing with microelectronics and nanotechnology related to telecommunications technology including semiconductors, it has a mission to undertake research and development meeting the needs that the industry will have in 3 to 10 years, or at least two generations ahead of manufacturing. It serves as a

[3] The factual data about IMEC in this chapter is correct as of year 2003, and is quoted from IMEC (2004) unless otherwise stated.

research provider that aggressively provides R&D programs in the semiconductor industry.[4]

In order to show that it is one of the new rising players in the intermediate domain, this chapter discusses three characteristics of IMEC: (1) independence, (2) horizontal partnerships among private businesses, and (3) vertical partnerships between industry and academia.

First, IMEC stands neither as a company nor as a university, but as an independent organization. In its inaugural stage, two thirds of its R&D budget was funded by subsidiaries of the Flemish government (Asakura 2003), but it subsequently increased its independent funding by establishing research tie-ups with external firms and institutions (see Fig. 7A). In 2002, more than three quarters of its budget was self-funded. As seen in Fig. 7B, a large portion of its independent funds was from international businesses and organizations. Section 8.1.1 noted that cutting-edge research projects and technologies are dispersed throughout the world. This situation underlines the importance of the international distribution of financial resources for forming flexible R&D partnerships.

Second, IMEC has been entering into a very wide spectrum of horizontal alliances. This broadness is especially notable in the area of international diversity. A mere glance at its membership list reveals leading-edge semiconductor companies from around the world, including companies from Europe, the United States, Japan, South Korea, Taiwan, and Singapore, as participating members.[5] It is noteworthy that a growing number of global businesses have been signing up for its joint research programs under the name of "IMEC Industrial Affiliation Program" or "IIAP," especially in recent years, as portrayed in Table 2. Moreover, the magnitude of its partnerships is also outstanding. The number of companies and institutions taking part in joint

[4] Based on the presentation of Professor Declerck, IMEC CEO, on June 13, 2003 at the University of Tokyo.

[5] This is according to the presentations of Professor Declerck and Professor Keersmaecker on June 13, 2003 at the University of Tokyo. IMEC has partnerships not only with individual companies but also with semiconductor-related consortiums in Japan and in the United States, such as US-based ISMT and Japan-based Selete.

research schemes with IMEC exceeds 500. It has ties not only with global businesses, but with start-up companies as well. It has the potential to form links not only among large firms, but also between large firms and new businesses.

**Fig. 7.A** Trends in self-generated income, **B** Subdivision of income from contracted research

A

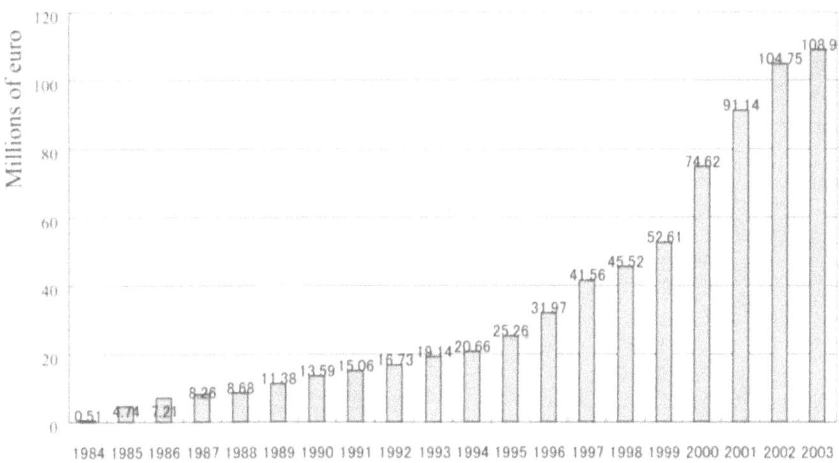


B

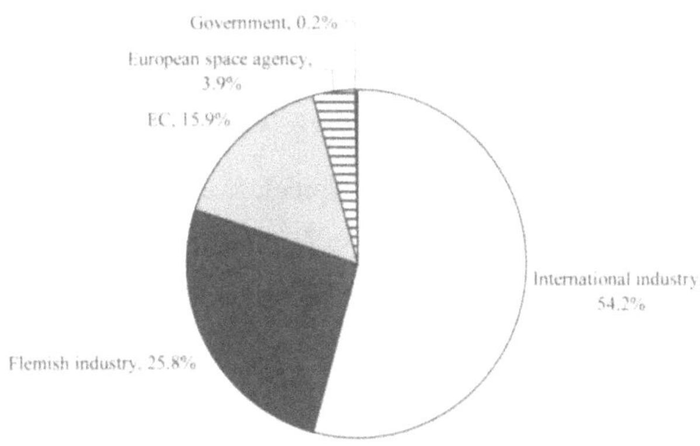


Source: prepared from IMEC (2004)

**Table 2.** Variation in participation of global enterprises in the IMEC Industrial Affiliation Program (IIAP)

| | IIAP area | | | | | | |
|---|---|---|---|---|---|---|---|
| Participating company | Lithography | High k | Interconnects | USJ/sllic | Cleaning | Emerald | Ge |
| Philips | + * | * | * | * | + * | * | * |
| Infineon | + * | * | * | * | + * | * | * |
| Texas Instruments | + * | * | * | * | + | * | |
| Intel | + * | * | * | * | * | * | * |
| AMD | + * | * | * | | + | | |
| Samsung | + * | * | * | | * | * | * |
| STM | * | * | * | * | * | * | * |
| Motorola | + * | * | | * | | | |
| Matsushita | | * | | | | | |
| Renesas | | * | | | | | |
| Micron | + * | | | | | | |
| National | + | | | + | + | | |
| Sony | | | + | | | | |
| Cypress | + | | | | | | |
| Via US-based ISMT only | | | | | | | |
| IBM | * | * | | | | | |
| HP | * | * | | | | | |
| Agere | * | * | | | | | |
| TSMC | * | * | | | | | |

The plus sign represents participation in 1999 when a cumulative total of 17 companies participated in the IIAP. The asterisk represents participation in 2003 when a cumulative total of 56 companies participated in the IIAP. The participating companies increased by 229 % in terms of the cumulative total number. Source: prepared on the basis of Van den Hove (2003)

IMEC is also unique because of its system for building horizontal partnerships. This is the IIAP (see Fig. 8). Jointly run by IMEC's own researchers, as well as researchers participating from external companies, a single IIAP program may involve multiple firms. One of the distinctive features of the program lies in the management of intellectual property rights. The results of joint research are split into two parts: the basic research component (in Pasteur's Quadrant) and the pure applied research part. Intellectual property rights in the first category are co-owned by IMEC and participating companies, while those in the second category fall under the exclusive ownership of the respective industrial partners. The intellectual property rights classified as $R1_B$ are shared by IMEC and partner B, while those of $R2_B$ are exclusively possessed by partner B. If any other company wishes to use the $R1_B$ results, it has only to conclude a contract with IMEC alone. This intellectual property right management system is designed to eliminate the necessity to separately contract with multiple individual companies for the use of each set of results.

**Fig. 8.** Intellectual property (IP) model of IMEC's industrial affiliation program

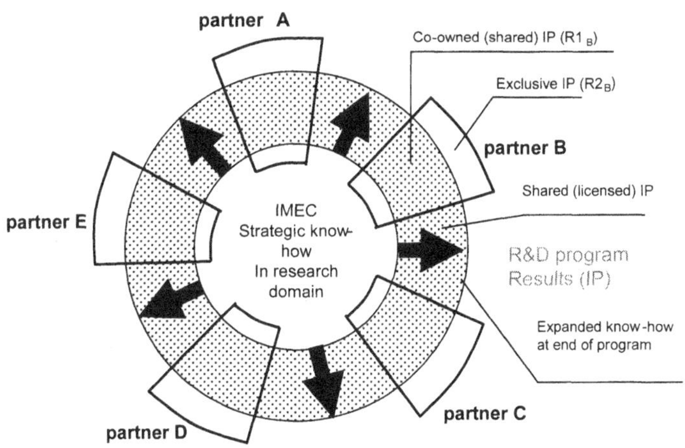


Reprinted from IMEC (2004)

The scheme has another unparalleled feature that cannot be found in conventional R&D partnerships among businesses: the IIAP's function of accumulating R&D achievements. The recent trend of R&D partnerships involves an increase in ad hoc coalitions based on contracts. This is attributable to the need to ensure flexibility. As a result of these ties, research findings are merely pooled in individual businesses. With ad hoc partnerships, it is difficult to see accumulated achievements finding their way to third parties. By acting as an independent third-party institution, IMEC carries out its own research, accumulates the results of R&D partnerships and offers them using the simple procedures described above. In so doing, it enables the partnerships to expand further.

Third, IMEC has close ties with universities forming vertical partnerships. The institution's research and development have a wider scope than that of International SEMATECH (ISMT), which is an American semiconductor R&D partnership. While the ISMT mainly covers the next generation, IMEC covers up to two or three future generations. Hence, IMEC's R&D activities are profoundly linked with pure basic research in the public domain. At the same time, it runs collaborative research with private businesses that are concerned with the private domain. It is naturally in a position to construct vertical partnerships between the public and private domains.

In addition, its mission, targeting not only the industrial world but also academia, drives the organization to bolster microelectronics research potential in Flemish universities. It therefore has cooperative relationships with six Flemish universities and runs joint research schemes with five associated laboratories in a broad array of fields. Based on these partnerships, some research programs that were launched by cooperative research organizations at universities are transferred to IMEC to make them into larger-scale study projects.[7]

These close R&D partnerships with public educational institutions allow research results to be additionally stored as IMEC's intellectual property, so that they can be used for applied research and development done by private businesses. IMEC acts as a research provider in the intermediate domain offering research programs that attract the participation of industrial partners, thus meeting the need for fresh approaches appropriate to the field of nanotechnology. If necessary, it transfers results of university research to link the public domain with the private one.

---

[7] For example, the Metal Organic Chemical Vapor Deposition (MOCVD) project was originally a R&D project run by the Department of Information Technology (INTEC) of Ghent University (RUG). In light of its strategic importance, it was transferred to IMEC in 2002 (IMEC 2003).

## 8.3 Research Providers

### 8.3.1 Past Successes and a Comparison with IMEC

Past successful R&D partnerships in the semiconductor industry include the VLSI Technology Research Association in Japan (1976–1979) and the former SEMATECH[8] in the United States (launched in 1987). By comparison with these successful attempts, this chapter reviews the three key features of IMEC as a research provider.[9]

In terms of independence, both the VLSI association and the former SEMATECH stood not as embodiments of contractual relationships among participating firms, but as independent legal entities. In this respect, they were independent organizations. However, they were both established on the initiative of their respective national governments. Also, they were largely dependent on government financial support and subject to the restrictions imposed by national policies. They stand in clear contrast with IMEC, which is independent in fact, as well as in theory.

With respect to horizontal partnerships, both the VLSI association and the former SEMATECH were involved in some partnerships of this kind linking private businesses. In both cases, however, the partnerships were aimed at increasing the competitiveness of the industry in the respective nations and all benefits were therefore accessible only to enterprises in each country. Each of the past two schemes that thrived created partnerships to deepen ties among companies within the country, but unlike those established by IMEC, they did not create new relationships on a global scale. Like the Belgian organization, they acted as a nodal point for creating partnerships, but they differed from IMEC in

[8] In the mid-1990s, SEMATECH achieved independence from governmental funds allowing it to accept non-American companies as members. In 2000, it was renamed "International SEMATECH." In view of these changes, when talking about the period before the mid-1990s this chapter refers to it as the "former SEMATECH," to distinguish it from the current consortium.

[9] Refer to Tanaka (2003) for further details on the VLSI association and the former SEMATECH.

one aspect: they acted as a "bond" among private companies within their respective countries while IMEC acts as a "bridge" to new global relationships.

As for vertical partnerships, the two past successful cases actually did not have close ties with universities. The VLSI association had had a number of original ideas before the project started and it was mainly engaged in evaluation and application on the basis of these ideas (Sakakibara 1986). It had no need to deepen its relationship with pure basic research in the public domain. For the former SEMATECH, the core challenge was harmonizing interfaces between the semiconductor device industry and the semiconductor manufacturing industry. In addition, it did not necessarily function as a nodal point because Semiconductor Research Corporation (SRC, 1982–) was there to serve as a channel for offering funds from the semiconductor industry to universities. In contrast with these facts, the industry is now facing a challenge in nanoscale technologies and is seeking an entirely different academic approach. It therefore needs a nodal point like IMEC that serves as a link to vertical partnerships.

### 8.3.2 Functions of the Research Provider

On the basis of the above speculations, the functions of a research provider are now summarized. First, it must exist as an independent body in the intermediate domain. To access cutting-edge R&D resources around the globe and to form flexible partnerships, it needs to have the freedom to act on its own initiative without being bound by government policies and corporate interests.

Second, it constructs horizontal partnerships among private businesses. When it does so, it needs to have an open relationship-building process so that new interbusiness relationships can be generated instead of further tightening limited coalitions among participating companies. To this end, a possible solution is to draw a distinction between private ownership and collective ownership of R&D outcomes, thus making it easier to access shared achievements. It is also important for the research provider to accumulate R&D results, because this opens the way for sharing contexts with partners and for bringing them into the next set of R&D activities.

Third, the research provider acts as a link for setting up vertical partnerships between the public and private domains. By conducting research and development in Pasteur's Quadrant on its own, it bridges the gap between the public and private domains to bring them together. As explained above, the research provider proactively offers R&D programs in the intermediate domain to enable a composite tie-up between horizontal and vertical R&D partnerships.

## 8.4 Conclusion and Tasks for the Future

This chapter started with a review of R&D partnerships from the two perspectives of intercorporate relationships and industry–university collaboration to explain how the intermediate domain, which lies between the private and public domains, combines these two types of partnership. It then applied this paradigm to R&D partnerships in the semiconductor industry to demonstrate that there is an independent bridge, under the name of "research provider," which functions in the intermediate domain.

One research provider, IMEC, was examined as an example of compound R&D partnerships based on the intermediate domain. We can now point to some topics for future research. The first is an investigation into what innovation will ensure that research providers function effectively. This must be studied in light of varying patterns of innovation even within a single industry. The second challenge is partnerships within the intermediate domain. In the semiconductor industry, there is an alliance of organizations including IMEC in Europe, International SEMATECH in the United States, and Semiconductor Leading Edge Technologies (known as "Selete") in Japan. This type of partnership should also be considered. The third challenge is a review of national governmental research institutions. Although this chapter does not cover these institutions, some of them, such as Japan's MIRAI Project and France's Minatec Project, play a central role in research and development involving nanoscale semiconductors. The fourth challenge is the relationship with the local economy. Through its partnerships with small and midsize businesses within the Flemish region, and by helping to establish numerous business ventures, IMEC, the organization that we have examined in this chapter, is closely linked with the local economy. The relationship with the local economy must also be scrutinized from the viewpoint of the compound nature of partnerships.

## Acknowledgments

This study was supported by the A06 Program Research (with Professor Osamu Sudoh of the University of Tokyo as its representative) in the "Informatics Studies for the Foundation of IT Evolution," as a specific-area study funded by a subsidy for scientific research from the Government of Japan. The author expresses his gratitude to all those involved. Also, the author deeply appreciates the assistance with information about IMEC's activities from the following IMEC officials: Professor Gilbert Declerck, President and CEO; Professor Roger De Keersmaecker, Vice President; Dr. Van den Hove, Vice President; and Dr. Akihiko Ishitani, Representative of Japan. Thanks are also offered to anonymous reviewers for their helpful suggestions.

## References

Arrow KJ (1962) Economic welfare and the allocation of resources for invention. In: The rate and direction of inventive activity: economic and social factors. Princeton University Press, Princeton, pp 609–625

Asakura H (2003) European Research and Development Consortium: companies as partners in contrast to the perception in Japan (in Japanese). Nikkei Micro Device, May 2003:119–121

Bressand A (1991) Net world: world vision in the twenty-first century (in Japanese). Toyo Keizai, Tokyo

Bush V (1945) Science: the endless frontier. A Report to the President by Vannevar Bush, Director of the Office Scientific Research and Development

Cohen WM, Levinthal DA (1990) Absorptive capacity: a new perspective on learning and innovation. Administrative Science Quarterly 35:128–152

Cohen WM, Nelson RR, Walsh JP (2002) Links and impacts: the influence of public research on industrial R&D. Management Science 48:1–23

Council for Science and Technology Policy (2002) Basic principles and policies for promotion of collaboration among industry, academia and government (in Japanese). http://www8.cao.go.jp/cstp/output/iken020619_3.pdf. Cited on February 24, 2005.

D'Aspremont C, Jacquemin A (1988) Cooperative and noncooperative R&D in duopoly with spillovers. The American Economic Review 78:1133–1137

Doz YL, Hamel G (1998) Alliance advantage: the art of creating value through partnering. Harvard Business School Press, Boston

Edler J, Meyer-Krhmer F, Reger G (2002) Changes in the strategic management of technology: results of a global benchmarking study. R&D Management 32:149–164

Florida R (1997) The globalization of R&D: results of a survey of foreign-affiliated R&D laboratories in the USA. Research Policy 26:85–103

Freeman C (1991) Networks of innovators: a synthesis of research issues. Research Policy 20:499–514

Fusfeld HI, Haklisch CS (1985) Cooperative R&D for competitors. Harvard Business Review Nov/Dec 1985:60–76

Gomes-Casseres B (1996) The alliance revolution: the new shape of business rivalry. Harvard University Press, Cambridge, MA

Hagedoorn J (2002) Inter-firm R&D partnerships: an overview of major trends and patterns since 1960. Research Policy 31:477–492

Hounshell DA (1996) The evolution of industrial research in the United States. In: Rosenbloom RS, Spencer WJ (eds) Engines of innovation: U.S. industrial research at the end of an era. Harvard Business School Press, Boston, pp 13–85

IMEC (2003) Annual report 2002: seeds for tomorrow's world. IMEC, Leuven

IMEC (2004) Annual report 2003: seeds for tomorrow's world. IMEC, Leuven

Link AN, Vonortas NS (2000) Research partnerships. Research Policy 29:567–586

Leopold G, Lammers D (2000) U.S. official raps German role in lithography group. EE Times, June 2, 2000

Ministry of Economy, Trade and Industry, Committee of Semiconductor Industry Strategy Promotion (2002) Challenges and policies of the semiconductor industry in Japan (in Japanese). http://www.meti.go.jp/kohosys/press/0002702/0/020508handotai.pdf. Cited on February 24, 2005

Mowery DC (2003) Using cooperative research and development agreements as S&T indicators: what do we have and what would we like? Technology Analysis and Strategic Management 15:189–205

Nishi Y (2003) Japan's nano-tech research not linked to competitiveness (in Japanese). Nikkei Micro Device, May 2003:97–104

OECD (2002) OECD science, technology and industry outlook. OECD, Paris

Roberts EB (2001) Benchmarking global strategic management of technology. Research・Technology Management, March–April, 2001:25–36

Rosenberg N, Nelson RR (1994) American universities and technical advantage in industry. Research Policy 23:323–348

Rosenbloom RS, Spencer WJ (1996) Technology's vanishing wellspring. In: Rosenbloom RS, Spencer WJ (eds) Engines of innovation: U.S. industrial research at the end of an era. Harvard Business School Press, Boston, pp 1–9

Sakakibara K (1986) Organization and innovation (in Japanese). In: Hitotsubashi Thesis Collection 86:160–175

Sanchez R (1995) Strategic flexibility in product competition. Strategic Management Journal 16:135–159

Sandholtz W (1992) High-tech Europe: the politics of international cooperation. University of California Press, Berkeley

Stokes DE (1997) Pasteur's quadrant: basic science and technological innovation. Brookings Institution Press, Washington DC

Sudoh O (1995) Compound network society (in Japanese). Yuhikaku, Tokyo

Tanaka H (2003) Joint research and evolution of the industrial system: functions of joint research in the semiconductor industry (in Japanese). The Bulletin of the Institute of Socio-Information and Communication Studies, the University of Tokyo 65:1–35

Teece DJ (1992) Competition, cooperation, and innovation: organizational arrangements for regimes of rapid technological progress. Journal of Economic Behavior and Organization 18:1–25

Teece DJ (1996) Strategic alliances and industrial research. In: Rosenbloom RS, Spencer WJ (eds) Engines of innovation: U.S. industrial research at the end of an era. Harvard Business School Press, Boston, pp 111–129

Van den hove L (2003) Sub-45 nm research programs in IMEC's 200 nm and 300 nm process research platform. A Presentation in IMEC Tokyo Executive Seminar, November 4 2003

Vernon R (1966) International investment and international trade in the product cycle. Quarterly Journal of Economics 80:190–207

Yoshimi M, Fukushima T (2002) Report on international technology roadmap for semiconductors (ITRS 2001) (in Japanese). Oyo Buturi 71:1241–1246

# 9 Regional Management and e-Democracy in the Information Society: Communications among Citizens and Officials on the Website Provided by Local Governments for Good Local Management

Ichiro Shinkai[i] and Koichi Naito[ii]

i NTT DATA Corporation, Toyosu Center Bldg., 3-3, Toyosu 3-Chome, Koto-ku, Tokyo 135-6033, Japan, shinkaii@nttdata.co.jp

ii NTT DATA Corporation, Kayabacho Tower Bldg., 21-2, Shinkawa 1-chome, Chuo-ku, Tokyo 104-0033, Japan, naitouk@nttdata.co.jp

**Summary**

As power devolves from central governments to local governments, the approach to regional management is now in question. By the enactment of the collective decentralization law local governments were certainly given the authority they need for proactive regional management, but given local fiscal restraints it is impossible to provide government that truly responds to the needs of citizens without a mechanism for regional management by collaboration between the government and its citizens. In many parts of the country there are examples of efforts to use e-Government, presently being constructed not merely as a tool for bringing government on-line and for providing the public with information, but also as a platform for addressing regional questions with the cooperation of governments and citizens. Electronic forums are often used as a tool for this purpose. This chapter examines closely some advanced case studies to assess the potential of e-Government for supporting the partnership between governments and citizens.

**Key Words** Regional management, Local government, e-Democracy, Citizen participation, Electronic forum

## 9.1 The Era of Regional Management

### 9.1.1 From an Industrial Society to an Information Society

Professor Iwao Kaneyasu characterized the industrial society as a goal-driven society (Kaneyasu 2002). So what was the goal for regional administration in the industrial society? Simply put, it was to eliminate regional gaps and achieve a national minimum standard of living, as defined by the central government with a view to ensuring balanced development nationwide. Under this system, both the authority and the financial resources were concentrated in the national government while prefectural and municipal governments could only serve primarily as organizations to control and check the implementation of policies adopted at the national level. This was despite the fact that prefectural and municipal governments were originally set up as entities for self-rule.

However, once the national minimum standard of living is achieved, the development-oriented administration of the industrial society is now no longer appropriate. Each local government must individually set its own goals on what to build on the established national minimum standard of living in accordance of its regional peculiarities. For this purpose, local public entities must have the authority to make decisions and to put them into practice.

It is under these circumstances that the collective decentralization law has been enacted. Decisions made by local governments rest with citizens and local governments define the targets of their policies by communicating with citizens. Today, with society's values so diversified, clear goals cannot always be set immediately. Regional governments find and define their targets by cultivating understanding and empathy between government and its citizens and among citizens. This is done through communication. Governments consequently are confronted with the challenge of changing the way they do things.

### 9.1.2 A Change Originating at the Local Level

Naturally, this situation is seen as a turning point, with the era of regional development in compliance with national-level plans coming to an end and a new age of regional management by local governments in collaboration with citizens beginning. Witnessing the cookie-cutter public works projects instigated by Tokyo and the environmental destruction they have wreaked, along with the deterioration in local public finances, local citizens have strong doubts about the current situation. This doubt has led to a gradual change in society. The change is first seen in areas where there is some contact with citizens. Local governments are ahead of the national government in reforming traditional administration, as in information disclosure, protection of personal information, and evaluation.

Broadly speaking, there are two scenarios for the changes taking place in local governments. In the first scenario, the head of a local government, who is seen as a sort of "reformer," senses a skepticism about the conventional procedures of which even the citizens themselves may not be clearly aware, and then exercises strong leadership to reform the autonomous body. Under the second scenario, government staffers that are dubious of the conventional practices make grassroots efforts to reform them. Of course, the first scenario requires the existence of personnel who understand the policies of their leader, while the second one demands that the head of the government and senior officials favor the actions taken by the staffers. But essentially, some local governments take a top-down approach, while others use a bottom-up technique. This chapter assesses these two methods by considering some actual cases.

### 9.1.3 Growth in Citizen Participation

As noted above, a growing sense of doubt among citizens about conventional local administration leads to change in local governance. This in turn means that in the future, regional governments will be increasingly expected to engage in a dialog with their citizens. The question thus arises: to what extent is the public actually motivated to take part in activities for their local communities?

In 2001 and 2003, NTT Data Corporation conducted a survey of individuals across the country, asking them about experience and the intention of participating in drafting municipal policies or organizing events for local development planned by municipal governments (Research Institute for System Science, Research and Development Headquarters, NTT Data Corporation 2002).[1] In the survey of 2001, 10.8% of the respondents had both experience and the intention of taking part in such activities. Another 4.5% had experience, but no intention of taking further part in such activities. Another 30.5% had the intention of taking part in such activities, although they had yet to do so. The responses to the same questions in the survey of 2003 were 20.7%, 9.6%, and 33.8%, respectively (see Fig. 1). In the intervening 2 years, the experience rate increased by 15% from 15.3% to 30.3%, and the participation intention rate increased by 13.2% from 41.3% to 54.5%. These figures indicate that citizen participation in local governance is increasing.

In drafting master plans for urban projects and the environment, those local governments that are positive about citizen participation not only provide opportunities for public hearings, but also organize study committees and workshops in which successful applicants from the public take part. These local public entities tend to face a problem: as more programs involving citizen participation are offered, all tend to have the same participants and they begin to tire of taking part. Local governments seek to diversify participation, but actual participation is drawn from a mere 20% of the population.

However, there is another group that accounts for 30% of the population. This is the group of people who are interested in taking part in activities, although they have never done so. NTT DATA labeled this group as the "potential group of citi-

[1] The results for the 2003 survey will soon be summarized on the same website given for 2001 (Research Institute for System Science, Research and Development Headquarters, NTT Data Corporation 2002).

zen participants." To establish regional management based on citizen participation, the key challenge will be to transform this group into key players and encourage them to actually participate.

**Fig. 1.** Experience and intention of participating in local policymaking or management and local promotional events

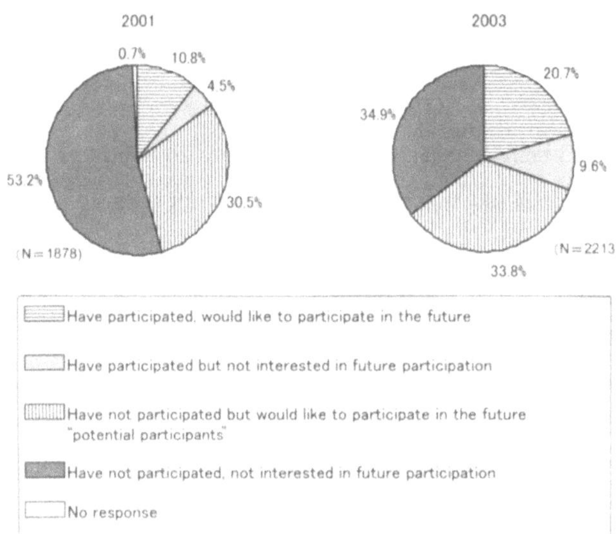


## 9.2 Achieving e-Democracy

### 9.2.1 The Group of Potential Citizen Participants and Information and Communication Technology

What opportunities should be given to highly motivated citizens? Even those local governments at the leading edge in terms of their position on involving the public in drafting master plans for urban projects and the environment had to address this question before they could encourage people to join the process. Mitaka City Government is proud of its long history of citizen participation, including the Mitaka Citizen Plan 21 Conference. The conference is composed entirely of citizens who applied for membership in 1999 to develop a basic plan literally from nothing without a draft. Even for this municipality, the development of citizen participation has been a history of steadily deepening involvement of local residents (Kiyohara 2000).

A future challenge will be the achievement of more extensive participation, instead of the current involvement of just a limited number of highly motivated people. To meet the target, participation channels should be provided by information and communication technology (ICT) to transcend time and space.

On the other hand, it is the potential participants comprising 30% of the population who are expected to join local management activities. Whether ICT is an effective tool to this group is an important question. From the above survey, information about the potential participants was revealed. In the survey of 2003, many of them were users of both conventional and cellular Internet services. In the survey of 2001, interest in taking part in communications with governments on the Internet was double the total average ratio of respondents (see Fig. 2). This result is strong evidence that the new participation channels provided by ICT are likely to encourage people in the potential group to actually get involved.

**Fig. 2.** Intention to participate in public policymaking on the Internet ($n$ = 2036).

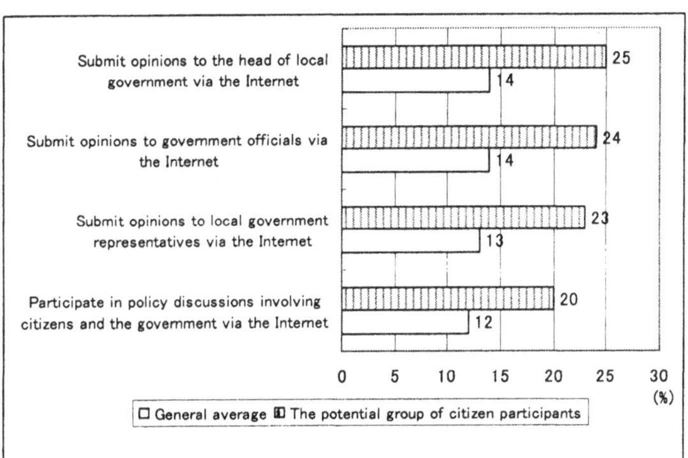


Source: NTT Data Corporation, "Individual Survey on e-Democracy in Japan" in December 2001

### 9.2.2 What Is e-Democracy?

Even in the days before personal computer (PC) Internet communications, moves were afoot to use ICT as a tool for participation in politics and administration. It was known by various names, including "digital democracy" and "cyber democracy," but this chapter uses the term of "e-Democracy" and defines it as "participation by citizens in the processes of policy formation, decision making, and implementation through the use of ICT." Participation in these processes can also be achieved by electing representatives such as members of local assemblies and heads of local government, by disclosing information about administration and assemblies, through discussions and consultations with governments and assemblies and among citizens, and in the form of direct participation in decision making as in referendums. In other words, e-Democracy has a number of possible styles.

Today, public attention is once again turning to the use of ICT for political and administrative involvement. Behind this situation lies the spread of the Internet. After some candidates took advantage of the Internet for successful election campaigns, the Internet became a popular part of the American political landscape (Yokoe 2001). Not confined to election purposes, communications between lawmakers and the electorate characterize e-Democracy in the United States.

In Japan, the public office election law restricts the use of the Internet for election campaigns and few members of the national Diet and local assemblies have used it for their political activities. Japanese e-Democracy is well developed in the field of public administration. It started with the construction of electronic government for extensive disclosure of administrative information, although this may not be true of Japan alone. With council minutes published on the website for the Japanese prime minister and the Cabinet, government transparency has been significantly increased. Other central government bodies invite public comments on their respective websites, but local governments, especially fundamental ones, are more enthusiastic about having more contact with the public. Within their limited budgets, they must now meet the diverse demands of their citizens. They realize that they alone cannot satisfy all of these demands. Many local public entities call for citizen participation to provide what citizens really want in the way they want it, and thus increase public satisfaction. These attempts include a policy for promoting collaboration with nonprofit organizations and the use of ICT. We turn now to consider some case studies involving the use of ICT.

### 9.2.3 Case Studies in Japan

Collection of public opinions through questionnaires and on the Internet, such as inviting the public to email the mayor, is a typical approach for encouraging citizen participation through ICT. An increasing number of local governments are launching electronic forums, because they allow citizens not merely to give opinions to administrative bodies, but also to discuss local issues with governments and with other citizens. A joint survey conducted by NTT Data and Keio Research Institute at SFC has confirmed that 733 local governments have electronic forums in operation all over the country (Keio Research Institute at SFC and Research Institute for System Science, Research and Development Headquarters, NTT Data Corporation 2003). Among these, this chapter focuses on the municipal governments of the cities of Fujisawa and Yamato, the both of which have been engaging in advanced activities, as well as the Mie Prefectural Government as an example of a broader local authority. In so doing, the different issues involved in e-forums run by local governments are considered.

#### *9.2.3.1 Fujisawa City (Approximate Population 380 000)*

The city of Fujisawa is located 50 min away from Tokyo and 20 min from Yokohama by train. The area is home to several universities such as the Fujisawa Campus of Keio University and the Shonan Institute of Technology. The city also has an industrial zone in its northern part. However, with nearly 80% of the city covered by residential districts, it is effectively a commuter town for workers and students in Tokyo and Yokohama. It also has Japan's most popular swimming beach and is well known as a tourist destination adjacent to Kamakura.

The Fujisawa City Citizens Electronic Forums have been in full operation since April 2001, after a preliminary trial in February 1997. The forums are the first in Japan to be launched by local governments for open discussions on all issues relating to municipal politics and are therefore a model for electronic forums in Japan. They are unique in that they are divided into two areas. The citizens' area is designed to foster community activity, while the city government area is expected to serve as a channel for citizen participation in municipal governance. The entire program is run by an operations committee consisting of citizens selected from applicants.

Individual forums in the citizens' area are hosted by those who launched them, while those in the city government area are run by people selected by the operations committee. In addition, the forums have advisors, professionals skilled in managing on-line communities, who are contracted to provide support for the operation by monitoring postings in forums and by assisting forum hosts (Fujisawa City Government 2005).

Electronic forum operators are usually concerned about anonymous attacks and libel, and this harassment has actually forced some forums to close. It seems, however, that Fujisawa City's forums have suffered no serious problems to date. There have been some complaints, including concerns about careless postings that have contained personal information, but these issues were resolved before they escalated. This is apparently because of the successful collaboration among hosts, advisors, the operations committee, and the municipal government.

Another concern about e-forums is that no one will comment. As explained above, there are more than 700 on-line conferences running across the nation, and in fact, some of them are fairly dull with few comments posted. However, Fujisawa City Government enjoyed 13 862 hits and 1150 postings in the citizens' area and 3875 hits and 484 postings in the city government area in October 2004. It has earned high marks for its achievement in creating active on-line communities (Fujisawa City Government 2004a).

The factors for the success of e-forums in the city of Fujisawa have not yet been fully explained, although some possible reasons have been suggested. First, composed of volunteers from the public, the operations committee emphasizes civic volunteerism. Second, the citizens' area has a freedom policy to ensure that anyone who lives, works, or studies in the city is entitled to open a forum on any subject and that anyone without any special qualifications can post comments. Third, the program enjoys some enthusiastic citizens who can take advantage of this free environment and some advisors who support them.

These factors are by no means unique to Fujisawa. Hypothetically, two types of reactions will be vital for active discussions in e-forums. One is a reaction to separate postings. No one would feel like having their say if they expect nobody to reply to their comments. Given that no one is constantly logged on to on-line forums, it is important to promptly reply to a posting even if it is a mere expression of appreciation or a greeting. That is one of the tasks of forum hosts. In the early stages, there may only be an exchange between hosts and individual participants; however, communications among participants will gradually emerge. Characterized by the fact that a forum in the citizens' area is hosted by the person who opened it, the Fujisawa approach is considered effective in this sense.

Another key point is the reaction of the municipal government to citizens' arguments. Citizens will quickly lose their zeal for participation if the government just listened and did nothing. There is no difference between the physical participation of a public hearing and on-line participation. One of the key points in establishing citizen participation is to provide citizens with feedback on their comments, not only to let them know how they are viewed by the government, but also to make them aware of the significance of their commitment. This point has also been stressed by Dr. Stephen Coleman of the Hansard Society, which has been striving to promote e-Democracy in the United Kingdom (Shinkai et al. 2002), and it has been confirmed by the joint trial that NTT Data conducted with a nonprofit organization based in Yachiyo City, Yachiyo Oikos (Yachiyo Oikos and Research Institute for System Science, Research and Development Headquarters, NTT Data Corporation 2003).

Fujisawa City Government has developed implementation guidelines to institutionalize a mechanism for incorporating discussions in e-forums into its policies.

To date, certain proposals, suggestions, and reports on opinion exchange have been produced for the City Government. In contrast, a general concern has been expressed that e-forum debates alone are singled out for special treatment. This is indeed a fair statement given the current state known as the "digital divide," in which Internet users are still in the minority. The city holds civic activities called "Kurashi Machizukuri Kaigi" (Living and Urban Development Forums) in its 13 districts. Administered by some 20 steering committee members, they offer opportunities for policy proposals and suggestions to the city government (Fujisawa City Government 2004b). As we have seen, the system of citizen participation in Fujisawa City is unique in that it guarantees various styles, including virtual online involvement and physical participation.

#### *9.2.3.2 Yamato City (Approximate Population 220,000)*

We turn now to the electronic citizens' forum of "Dokodemo Community" (Ubiquitous Community) in the city of Yamato (Yamato City Government 2004a), a suburban community about 1 hour away from central Tokyo and around 20 min away from Yokohama by train. With more and more people working and studying in Tokyo or Yokohama, it increasingly serves as a commuter town.

In the Fujisawa City case study, we confirmed that one of the keys to success is a structure based on hosts, advisors and the operations committee, as well as their collaboration. On the other hand, Dokodemo Community has no particular position equivalent to a host. According to its rules, every contributor is asked to categorize their own comments and the type (such as "response" or "speculation") of subsequent comments to them is restricted depending on the category (Yamato City Government 2004b). This forum expects the participants to structure their discussions themselves, but this attempt seems not to attain its goal completely. The forum has a series of comments labeled "monologs," but many are in fact inspired by remarks made by others rather than being genuine isolated monologs. In a forum with an administrator, the administrator will intervene to construct a discussion, for instance in asking contributors some questions about their comments or asking them to confirm their meaning. The rule of comment classification is designed to prevent any discussion from being affected by the personality or skills of the hosts, but on the basis of past experience, it is necessary to review the categories and rules if the objective is to be achieved.

Despite this problem, the Dokodemo Community is considered important as a successful example, because of a mechanism that ensures that all government officials take part in it. This allows Yamato City's forum to function as an effective interface between the government and citizens. Most local governments run their own websites to provide administrative information for citizens. Normally, these websites act as an interface between government and the public, but they cannot go beyond simply providing information from different divisions of the government to the general public. Overcoming this challenge, Yamato City Government has created a link between its administrative staff and every citizen by introducing the Dokodemo Community.

A bureaucratic organization is essential for effective administrative services, and individual staff members are incorporated into the segmented structure. On the other hand, citizens each have different and complex views or demands, which do not necessarily fit perfectly with the segmented administrative organization. Yamato City Government attempts to ensure the full participation of its staff members to overcome any conflict between what is a horizontal structure and a vertical one.

Remarks made by the public in the Dokodemo Community are delivered to all staff members by email so that those staff members who find that a comment touches on their duties can respond to it. Personnel are requested to check for any unanswered comment. This allows contributors to see that their remarks are handled sincerely by the government and to feel satisfied with information relevant to them. The forum has no institutionalized host, but it can be said that all government staff members act as hosts. For the administration, it paves the way for providing information to those who want it, in the way they want it, and because the communications are disclosed to the public, they are shared by participants. In terms of a dialog with the government, the forum engenders very high citizen satisfaction.

Needless to say, this system requires great effort on the part of administrative staff. In addition, the government has been criticized for making an excessive commitment to the forum and that there are many other things that they should be doing. However, with the city of Yamato having a population of 220 000, this program should be seen as evidence that any local government of this size can achieve one-to-one communication with its citizens to address every single challenge in the city.

#### *9.2.3.3 Mie Prefecture (Approximate Population 1,860,000)*

Above, we have considered the activities of local governments, which are the closest form of government to citizens. Our final case study focuses on an example of a wide-area local public entity, Mie Prefectural Government.

Mie Prefecture is located to the east of Nara Prefecture and is on the east coast of the Kii Peninsula. With a coastline of more than 1000 km, including Ise Bay and the Pacific Ocean, it is home to the some of the leading fisheries in Japan. Geographically long north to south, it is proud of its variety of natural riches. It has two national parks, two quasi-national parks, and five prefectural natural parks. These parks collectively account for more than one third of the entire land mass of the prefecture.

Under the leadership of former prefectural governor Masayasu Kitagawa, the Mie Prefectural Government has promoted reforms in accordance with its basic principle of "prefectural administration with a priority on people." After first working to change the attitudes of government officials, and then structurally reforming the government itself, the Mie Prefectural Government launched a series of administrative system reforms, the general thrust of which was to completely restructure prefectural governance where it involved the citizens. As part of this latter set of reforms, the "e-Democracy Forums" were launched in May 2002, to offer a means by which prefectural residents could participate in the development of the local community (Mie Prefectural Government 2004). This program was first proposed to senior officials as the result of a policy research workshop, and was adopted as a project. Consisting of volunteer staff, the workshop had been set up for human resource development reform.

At the start of the e-Democracy Forums, a broad spectrum of intellectuals, including university professors, organizers of nonprofit organizations, and journalists, acted as hosts called "e-editors" to propose subjects and organize discussions. They played a role similar to the hosts of the city government area of Fujisawa City Citizens' Electronic Forums discussed above. The difference is that the forums were run not by an operation committee consisting of selected voluntary citizens, but by an editorial council of e-editors, intellectuals, and some prefectural officials. In fact, this approach was revised after about 1 year. Today, any participant living, working, or studying in the prefecture, as well as the prefectural government itself, is free to open a new forum and act as an e-editor of the forum they launch. In addition, a larger "e-Democracy Representatives Council" has replaced the editorial committee in operating the program (Tanaka 2004). The principles for launching and administering e-forums have also been shifted from the initiative of the prefectural authorities to citizen autonomy.

The conditions of the forums vary depending on the subject. According to Mikiya Tanaka, the project leader responsible for the e-Democracy Forums, some subjects prompt repeated questions and replies between enthusiastic citizens and prefectural personnel. Others have difficulty generating postings despite the efforts of e-editors, and some subjects attract many opinions that are so diverse the discussions look chaotic (Tanaka 2002). It is believed that these circumstances are true of e-forums run by other local governments. Although it depends on the degree to which the public and administrative staff are accustomed to debates in electronic forums, they will gradually learn what subjects are appropriate for online discussions as more and more channels of citizen participation open.

Tanaka also explains that the primary advantage of the scheme lies in the fact that it has given rise to direct communication between citizens and prefectural officials. The personnel in the prefectural government formerly had few opportunities for direct conversation with the public, but now have increasing opportunities to do so through e-Democracy Forums. Direct communications with real people instead of statistical data has enabled the staff to adopt the stance of putting the public first. In this sense, it can be said that the project has reformed the mindset of prefectural officials.

Even so, future debates should assess the significance of direct communication between citizens and prefectural governments as the system is being developed, to ensure that administrative services for citizens are provided by local authorities at the most grassroots level. The systems offer advantages in harmonizing perspectives between prefectural and municipal governments when it comes to "priority on people." But as power continues to devolve, wide-area local governments will have to confront their raison d'être once municipal authorities become capable of meeting the needs of the public directly. They will have to justify their presence. In view of this development, it will be interesting to see what roles will be played by systems like the e-Democracy Forums.

#### 9.2.3.4 Overseas Conditions

An earlier part of this chapter mentioned that ICT is widely used for communication between politicians and voters in the United States as well as for election campaigns. In regard to local administrations, some on-line debates are organized by nonprofit organizations such as the Minnesota e-Democracy. They have a considerable impact on regional policies. In contrast, there are few such discussions sponsored by public authorities. In the pre-Internet period, an electronic forum service launched on the Public Electronic Network (PEN) of the California city of Santa Monica attracted public attention as an innovative concept. It was actually used, but the service is no longer provided. Because of disruptions as well as the quality of discussions, people seem to have left the forum. If public authorities in the United States monitor comments and expressions prior to publication in a forum as a means of combating disruptions, they run the risk of violating the right to freedom of speech. That makes it difficult for public entities to run any e-forums on their own and thus avenues of electronic discussion in the United States are mainly operated by nonprofit organizations (Dutton 2002).

Public authorities in the United States think of e-Democracy as providing information and services on websites that are easily accessible to citizens. They put emphasis on user-friendliness in building their information systems and then they invite people to give comments on government activities. A typical example is a portal of public comments, Regulations.gov (US Environmental Protection Agency 2004).

In the United Kingdom, emphasis has recently been placed on collecting public views. In July 2002, the Blair government conducted a public hearing on promoting e-Democracy, with e-participation and e-voting as two pillars. A website for e-Democracy was set up to provide relevant information and there are some on-line discussions in electronic forums (UK Cabinet Office 2004a). The former government portal known as "UK Online" focused on direct communication between the government and the public. The present portal known as "Directgov," however, does not seem to have distinctive characters from the viewpoint of direct communication between the government and the public (UK Cabinet Office 2004b) .

In Germany, the Bundestag, or federal parliament, offers on-line forums (Deutscher Bundestag 2004). The forums provide information on several subjects, show comments by members of parliament from major political parties, and collect public opinions on broad areas. Although some British local councils operate electronic forums, few on-line forums are run by assemblies (The Scottish Parliament 2004).

On the other hand, a South Korean initiative is fairly ambitious. Guangnam-gu in Seoul has email-based referendums, communications conferences between the government and citizens using Internet chat technologies, and discussions on e-forums. Its government actively implements those programs deemed to have attracted a large number of supportive comments in on-line forums (Ko 2003). It is an attempt at direct democracy.

At the national level, the draft of "e-Korea Vision 2006: The Third Master Plan for Informatization Promotion (2002–2006)" explicitly includes a policy of "preparing a detailed system to promote the electronic participation of citizens in the policymaking process by systemizing on-line public forums (Ministry of Information and Communication, Republic of Korea 2002)." Japan's "e-Japan Strategy II" also refers to "constructing a society that allows extensive public participation" as one of the objectives of putting administrative services on-line, which has been defined as a priority challenge. However, it does not have any specific description about the modes of participation (IT Strategy Headquarters 2003). Compared with this, the South Korean program is remarkably innovative.

## 9.3 Challenges for the Future

### 9.3.1 Future Administrative Services to Be Offered by Local Governments

The trend toward citizen participation in the policy–decision process of local governments will stay intact. Governments will have to make efforts not only in collaboration with a limited number of motivated citizens, but also in expanding bases for participation. The above-mentioned case studies that looked at electronic forums are indeed among the initiatives in this area.

There is an extremely heavy burden in time and effort in holding intensive discussions in a real forum such as workshops, while on-line exchange of views with the public for policymaking faces its own difficulties. According to Takashi Kobayashi[2], the current electronic forum represents considerable progress in providing information, but it still fails to produce an effect in terms of developing and marshalling arguments. Somc people claim that it is only through discussions among a small number of particular people that arguments can lead to a consensus.

Fujisawa City Government uses its citizens' electronic forums in combination

[2] Then Deputy Chief of Information Policy Division of Yamato City Government involved in setting up and operating the Dokodemo Community, now Assistant Professor at Tokai University.

with tangible deliberative committees for formulating several plans. This is a highly thought-provoking effort to provide appropriate means of discussion among the public and a venue for decision making. However, it will have to take responsibility in explaining how the on-line debates have influenced the committee deliberations.

Notwithstanding these challenges, electronic forums have the potential to be more than mere outlets for expressing views, given that they enable their participants to listen to the opinions of others and receive reactions to their own comments on a real-time basis. A deliberative poll that was held on-line in the United States in 2002 revealed that on-line discussions allow participants to modify their stances on their own (Political Communication Lab, The Center for Deliberative Polling 2004). To ensure this function, the specifications of e-forum software will have to be reviewed, facilitators must be more skilled, and participants should have better developed capabilities for comprehension and expression in exchanging ideas on the Internet.

Local governments are now at the stage of developing a foundation for citizen participation in administrative decision making. Their next job is to take on the new challenge of building a mechanism for the public responsibility of citizens. If citizens just take part in the administrative policy-forming process to decide on local management policies and later leave the implementation to the government, there will be no real difference from the conventional demand-making participation. The next challenge is to create a system that allows the public to participate in regional management itself.

Yamato City's local electronic currency, called the "Local Values Exchange System," or "LOVES," is a pioneering attempt to create such a system. The local currency, which is only valid within a specific zone and among specific people (Morino 2002), is designed to maintain and revitalize the community and economy in the locality.

LOVES is an electronic regional currency system embedded on IC cards. It has been developed as an application of Yamato City Government's project for the "construction of an e-community with the participation of all residents" adopted as a "research project on IT-powered cities with the propagation of IC cards, etc.," funded by a supplementary budget of the Ministry of Economy, Trade and Industry in fiscal 2000. The service started in April 2002. A total of approximately 87 000 IC cards had been issued to some 35 000 households as of March 2003. LOVES values are now traded for access to public facilities and events offered by the city government, registration and requests for citizen volunteers, registration and exchange of disused goods, as well as for the registration and use of discount services provided by shopping malls (Kobayashi 2002).[3]

While the Yamato City Government runs the Dokodemo Community as an attempt at local participation through discussions, it positions LOVES as a means for local participation through action. There are various ways to become involved in local communities. The Yamato City Government offers a mechanism for achieving different styles of participation. The authors believe that LOVES will possibly link on-line activities with the real world.

It is not at all easy to translate into reality some of the on-line opinions that have been proposed or approved. The administration is subject to budgetary and institutional limitations. Not every debate participant has the time and energy to take part in activities. Electronic forums are originally designed to provide a new participation channel for those who cannot participate in local activities for reasons of time and place. However, if nothing is done, on-line debates will be confined to the virtual world and will not be linked with social activities. Within those confines, they will have limited impacts on the real world. The authors see the Yamato City Government as trying to connect the virtual world with the real world by providing in LOVES a means of exchange that is different from conventional currency, as a means of transcending these limitations.

[3] The number of cards issued and other details can be found at http://www.city.yamato.kanagawa.jp/Jyoho/award/jasi.htm, last cited on 5 November 2004

### 9.3.2 A New Mechanism of Checks and Balances

This chapter has discussed direct participation in administration. We should also study its impact on representative democracy. As a matter of course, the administration is not a decision-making body of a local public entity. Both its budgets and its ordinances are decided by the assembly. However, the reality of the assembly is that most bills are submitted from the head of the government. If the bills are prepared through discussions directly involving citizens, the assembly will have no option but to respect the public opinions they encapsulate and may no longer do anything but offer an endorsement. If the trend toward citizen participation in administration is irreversible, the assembly must identify the role it will play in this era on its own, instead of blaming the direct participation for disregard of legislative power.

One possible solution would be to define policy priorities from a general perspective in view of other issues within the budgetary limitations of local governments. Even though citizen participation is widened, actual participants in administrative activities including those using ICT will probably remain limited to a small number. It should be an "argument of the haves" that the blame for the disadvantages of nonparticipation must be borne by those citizens who do not participate although they have the opportunity to do so. A system such as electronic forums is sometimes seen as a tool for hearing silent voices, but no one can have their say without the necessary environment and capabilities. In other words, it is by no means possible to hear all voices perfectly. The government will naturally pay attention to the "silent voices" hidden in a broad array of information, but the assembly still has a great role to play as the body making the ultimate and comprehensive decisions.

Even so, the assembly has many hurdles to clear before it can claim to play such a role. First, it must achieve a certain level of turnout in its elections, or it will be unpersuasive when arguing that it consists of elected representatives of citizens. Assembly members should endeavor to raise the turnout, and they must also communicate with the electorate from all walks of life to constantly keep track of public opinion. If they merely represent the limited interest groups that vote for them in elections, the assembly can expect no growth in public trust. Despite some possible differences between the United States and Japan in the circumstances surrounding lawmakers, including the size of their support staffs, it is necessary for council members in Japan to make full use of ICT in strengthening their ties with a wide range of voters. These efforts will reestablish confidence in the assembly as a place where diverse views are expressed.

## Acknowledgments

In compiling this chapter, the authors have learned much from the debates in the Next-Generation Electronic Government Study Group, the Consensus Building Method Study Group, the e4p Study Group, and the Regional Media Strategy Study Group. The authors express appreciation to all those involved. Special thanks are due also to the officials of Fujisawa City, Yamato City, and Mie Prefectural Governments, who provided the authors with excellent explanations of their activities as well as valuable suggestions for the section on case studies. If this paper contains any inaccuracies or irrelevancies, the responsibility lies with the authors. Finally, the authors express their gratitude to those involved in the "Pillar A06" studies at the "Informatics Studies for the Foundation of IT Evolution," which was founded by a subsidy for scientific research from the Ministry of Education, Culture, Sports, Science and Technology, for offering a rare opportunity to make a presentation.

## References

Deutscher Bundestag (2004) http://www.bundestag.de/dialog/forum/index.htm. Cited 5 November 2004

Dutton WH (2002) Society on the line: information politics in the digital age (in Japanese). Tokai University Press, Hadano, Kanagawa

Fujisawa City Government (2004a) Forums in operation as of October 2004 (in Japanese). http://www.city.fujisawa.kanagawa.jp/~denshi/page100159.shtml. Cited 5 November 2004

Fujisawa City Government (2004b) Outline of the "Living and Urban Development Forums" system for citizens' proposals (in Japanese). http://www.city.fujisawa.kanagawa.jp/jiti/data05886.shtml. Cited 5 November 2004

Fujisawa City Government (2005) Overview of Fujisawa City citizens' electronic forums (in Japanese). http://www.city.fujisawa.kanagawa.jp/~denshi/page100095.shtml. Cited 2 March 2005

IT Strategy Headquarters (2003) e-Japan strategy II (in Japanese). http://www.kantei.go.jp/jp/singi/it2/kettei/030702ejapan.pdf. Cited 5 November 2004

Kaneyasu I (2002) A way of developing projects (in Japanese). Chuokoron, Tokyo

Keio Research Institute at SFC and Research Institute for System Science, Research and Development Headquarters, NTT Data Corporation (2003) Survey on installation of electronic civic forums (in Japanese). http://www.riss-net.jp/edemo/. Cited 5 November 2004

Kiyohara K (2000) A new era of local governments originating from Mitaka (in Japanese). Gyosei, Tokyo

Kobayashi T (2002) Local electronic currency of LOVES (in Japanese). http://web.sfc.keio.ac.jp/~yamatotk/netcom/loves-kobayashi.pdf. Cited 5 November 2004

Ko Y-S (2003) Progress and achievement in construction of e-government in the special city of Seoul (in Japanese). Gyosei and ADP 39:18–25

Mie Prefectural Government (2004) What are e-democracy forums? (in Japanese). http://www.e-demo.pref.mie.jp/how.php. Cited 5 November 2004

Ministry of Information and Communication, Republic of Korea (2002) e-Korea vision 2006: the third master plan for informatization promotion (2002–2006). p 42 (A summary Japanese translation of the provisional edition by Sawai T, available at http://japan.internet.com/public/materials/pdf/20021004/ekoreavision2006.pdf. Cited 5 November 2004

Morino E (2002) Local independence and the perspective of monetary reform theory (in Japanese). Kikan Mirai Keiei 7:38–43

Political Communication Lab, The Center for Deliberative Polling (2004) First online deliberative opinion poll reveals informed opinions on world problems.

http://pcl.stanford.edu/common/docs/research/fishkin/2003/onlinedpoll.pdf. Cited 5 November 2004

Research Institute for System Science, Research and Development Headquarters, NTT Data Corporation (2002) Individual survey on e-democracy in Japan. http://www.nttdata.co.jp/en/find/report/index.html for the summary. Cited 5 November 2004

Shinkai I, Kasuga M, Yamada E, Kanaya T (2002) Local strategies based on e-democracy (in Japanese). Shogakukan Square, pp 163–164

Tanaka M (2002) Will e-forums give rise to e-democracy? (in Japanese). Presentation material at the 4th Regional Media Strategy Workshop Meeting, Keio University, 18 December 2002

Tanaka M (2004) A case study in Mie Prefecture (in Japanese). In: Kaneyasu I, Nagasaka T, Shinkai I (eds) Guidelines on electronic civic forums. Gakuyo Shobo, Tokyo

The Scottish Parliament (2004) http://www.communitypeople.net/interactive/. Cited 5 November 2004

UK Cabinet Office (2004a) In the service of democracy. http://www.edemocracy.gov.uk/. Cited 5 November 2004

UK Cabinet Office (2004b) Directgov. http://www.direct.gov.uk/Homepage/fs/en. Cited 5 November 2004

US Environmental Protection Agency (2004) Regulations.gov. http://www.regulations.gov/. Cited 5 November 2004

Yachiyo Oikos and Research Institute for System Science, Research and Development Headquarters, NTT Data Corporation (2003) Report on experiment of Internet tags in Yachiyo City (in Japanese). NTT Data Corporation, Tokyo

Yamato City Government (2004a) Ubiquitous community (in Japanese). http://www2.city.yamato.kanagawa.jp/index.html. Cited 5 November 2004

Yamato City Government (2004b) Introduction to the Yamato City electronic information exchange system (in Japanese). http://www2.city.yamato.kanagawa.jp/dokodemoppt/ppframe.htm Cited 5 November 2004

Yokoe K (2001) E-Politics. Bungeishunju, Tokyo

# Index

K



L



M



N



O



P



R

FSC
www.fsc.org
MIX
Papier | Fördert
gute Waldnutzung
FSC® C083411

Kolibri 360 GmbH
Ferdinand-Jühlke-Straße 7
99095 Erfurt, Deutschland
produktsicherheit@kolibri360.de